Praise for *Useful Idiots*

"Mona Charen shows us in all its gory detail the history of the liberal Left's hatred of America, from toadying to murderous communist regimes, even during the Cold War, to excusing the slaughters committed today by Islamofascists. *Useful Idiots* reminds us that the West is endangered by enemies within who seek to weaken our morale even as we face assault from without."

—Robert H. Bork, senior fellow,
American Enterprise Institute

"Mona Charen documents the endorsement of Marxist-Leninist rhetoric by the American Left throughout the Cold War and the current attempt to hide that ugly legacy. All those interested in Cold War politics should add this to their library."

—William J. Bennett,
author of *Why We Fight*

"This valuable book documents the extent of liberal bias in American political commentary. It is simultaneously insightful, delightful, and deeply disturbing."

—Jeane Kirkpatrick,
former U.S. ambassador to the United Nations

Bill Fitz-Patrick

## *About the Author*

MONA CHAREN's column on politics and culture is syndicated in more than two hundred newspapers. Before becoming a columnist and television commentator on CNN's *Capital Gang,* Charen wrote speeches in the Reagan White House for Nancy Reagan and worked on the presidential campaign of Jack Kemp. After graduating from Barnard College, Columbia University, she began her career in journalism at *National Review* magazine. She lives in Virginia with her husband and three children.

# Useful Idiots

# Useful Idiots

How Liberals Got It Wrong in the
Cold War and Still Blame America First

## MONA CHAREN

Perennial

*An Imprint of* HarperCollins*Publishers*

First Perennial edition published 2004.

Library of Congress Cataloging-in-Publication Data is available.

ISBN 0-06-057941-2

04 05 06 07 08  ❖/RRD  10 9 8 7 6 5 4 3 2 1

*For my precious*
*Jonathan, David, and Benjamin,*
*who understand so much already*

# Contents

INTRODUCTION TO THE PERENNIAL EDITION                                    xi

INTRODUCTION:     None Dare Call It Victory                               1

CHAPTER ONE:      The Brief Interlude of Unanimity
                  on Communism                                           11

CHAPTER TWO:      The Consensus Unravels                                 23

CHAPTER THREE:    The Bloodbath                                          55

CHAPTER FOUR:     The Mother of All Communists:
                  American Liberals and Soviet Russia                    77

CHAPTER FIVE:     Fear and Trembling                                    119

CHAPTER SIX:      Each New Communist Is Different                       171

CHAPTER SEVEN:    Post-Communist Blues                                  231

EPILOGUE                                                                259


NOTES                                                                   265

ACKNOWLEDGMENTS                                                         287

INDEX                                                                   291

# Introduction to the Perennial Edition

IN MARCH 2003, the United States undertook to remove one of the world's most savage and dangerous dictators. Unlike other despots ruling over sandpiles in other places, Saddam Hussein was in possession of a large army and huge natural wealth in the form of oil and advanced weapons, including weapons of mass destruction. In the center of a region that had spawned a particularly vicious form of anti-Americanism, he had cozy ties with a variety of terrorists and extremists. Saddam had attacked no fewer than three countries, had committed genocide within Iraq, and harbored ambitions to dominate the Gulf region. In the wake of September 11, 2001, the Bush administration, with broad public support, decided that the United States could no longer rely on deterrence to shield us from threats simmering in far-off places. In an age when weapons that can kill thousands can be deployed by a few individuals willing to die for their cause, a figure like Saddam being in possession of such an arsenal represented an ongoing and ominous threat.

The people of the world ought to have thanked us. If we lived in a better world, they would have. As it was, the U.N. Security Council declined to support the use of military power to enforce its own resolutions, and the United States and Britain fielded brickbats from France and Germany. The street protesters came out in force, both in the United States and Europe. It was "blood for oil," they chanted, and the true "axis of evil" consisted of Bush, Cheney, and Rumsfeld. The leftists are always in such a state of apoplexy about America that it's fun to try to imagine where they could go (rhetorically) if the United States ever actually committed the sort of aggression of which they are always accusing us. Imagine if the United States invaded Venezuela for her oil fields! What could the Left say beyond what it's already said when we merely liberated Iraq from a systematic sadist?

One might have expected American liberals, who supposedly cherish liberty, the rule of law, and democracy, to recognize the strategic and moral case for taking out Saddam Hussein. A handful did. But mainstream liberal opinion slid reflexively into the distrust of American power and skepticism about America's interests that has characterized it since the Vietnam War.

It isn't that liberals are pacifists. They were enthusiastic backers of military force in Bosnia and Kosovo, two conflicts where no U.S. interests were implicit at all. This absence of U.S. interests seemed to purify the use of force in liberal eyes. The same spirit was in evidence when intervention in Liberia was first discussed. Many of those who just finished decrying the "rush" to war in Iraq were prepared to send U.S. soldiers to Liberia minutes after learning that the U.N.'s Kofi Annan had requested them. When the United States acts in a purely humanitarian fashion, liberals are ready to dispatch the marines. When the security of the United States is implicated, liberals insist upon diplomacy alone.

*Useful Idiots* is not just a book about how wrong liberals were about the Cold War—it is also an exploration of their hostility toward the United States. This hostility gets expressed in many ways. In wartime, it is most obvious in the tendency 1) to distrust America's stated motives for going to war, 2) to believe that America will not prevail, and 3) to embrace and amplify any error or misjudgment by U.S. policy makers as proof of bad faith.

The picture that most liberals carry in their mental wallets was formed during the Vietnam War. During that war, they became convinced that America was often on the "wrong" side of conflicts in the Third World; that our military was likely to be clumsy if not outright criminal; that the United States was an international bully; and that intervention in other nations' affairs was nearly always illegitimate.

Columnist and television host Chris Matthews predicted a year before the Iraq War that "this invasion of Iraq, if it goes off, will join the Bay of Pigs, Vietnam, Desert One, Beirut, and Somalia in the history of military catastrophe." Senator Edward Kennedy said, "The American people don't want this war. Our global allies don't want this war. So why is President Bush stampeding down the warpath, and not working toward a real solution to disarm Saddam?" The "stampede," if it were dated from September 11, 2001, took eighteen months. If it were dated from the close of the Gulf War in 1991, when Saddam first began to flout the truce terms, it took twelve years.

Senator John Kerry told a cheering crowd, "We should never fight a war because we want to. We should fight a war only because we have to." This was precisely the mind-set of the British and French during the 1930s when Hitler was militarizing the Rhineland and taking Czechoslovakia. If Britain and France had battled Germany in the early or mid-1930s, Hitler could not have won. But they stalled and appeased until the end of the decade, when Hitler was much stronger. Kerry's axiom is exactly false: It is a lucky nation that fights wars only when it wants to, not when it is forced to. In the case of the United

States, it goes without saying that we would never fight wars of aggression or conquest in the twenty-first century. If we do choose to fight it is because we are preventing a medium- or long-term threat from becoming imminent.

But most Democrats were with Kerry. They viewed the resort to force not as a necessary expedient in a dangerous world but inevitably as a failure of diplomacy or worse. Senate Minority Leader Tom Daschle told reporters on the eve of the Iraq War that he was "saddened that we have to give up one life because this president couldn't create the kind of diplomatic effort that was so critical for our country." Representative Dennis Kucinich, an Ohio Democrat, offered during the war, "We should get out. The appropriate action right now is to spare the lives of our men and women who serve, to spare the lives of innocent Iraqis, for the U.N. to come in, the U.S. to step back." Such a course would, of course, have kept Saddam and his fascistic regime in power, weakening the United States immeasurably in the eyes of friends and foes alike.

The liberal press weighed in heavily. "American war planners clearly misjudged the determination of Iraqi forces." So announced reporter Peter Arnett on Iraqi television less than two weeks after the Iraq War had begun. Arnett, whose earlier forays into America-bashing are detailed in Chapter Two, seemed determined to prove that nothing has changed. In his quite expansive discussion with Saddam's flunkies, Arnett praised the spirit he claimed to find in the Iraqi people (who would be under new management in less than one week). "This is clearly a city that is disciplined. The population is responsive to the government's requirements of discipline." Arnett then painted a contrast with the United States. "It is clear that within the United States there's a growing challenge to President Bush about the conduct of the war and also opposition to the war. So our reports about civilian casualties here, about the resistance of the Iraqi forces, are going back to the U.S. and it helps those who oppose the war."

Arnett was promptly fired from his job with NBC—he had let the mask slip a bit too obviously this time. He was openly acknowledging what had been only implicit in his coverage of the first Gulf War for CNN, namely that he was attempting, through his tendentious reporting, to help Saddam Hussein. But his assessment of the war's progress was not much different from that of most reporters covering the war for the major networks and newspapers. They were prepared, not to say primed, to interpret all news in the light least favorable to the United States.

When the war was only days old, the *New York Times* pronounced on page one that it had become a "quagmire." Scott Ritter, the onetime U.N. weapons inspector who changed sides and became an antiwar advocate, told South African TV in early April, "The United States is going to leave Iraq with its tail between its legs, defeated. It is a war we cannot win. We do not have the military means to take over Baghdad and for this reason I believe the defeat of the United States in this war is inevitable. Every time we confront Iraqi troops we may win some tactical battles, as we did for ten years in Vietnam, but we will not be able to win this war, which in my opinion is already lost."

Writing in the online magazine *Slate*, Robert Wright reflected, "As the war drags on, any stifled sympathy for the American invasion will tend to evaporate. As more civilians die and more Iraqis see their 'resistance' hailed across the Arab world as a watershed in the struggle against Western imperialism, the traditionally despised Saddam could gain appreciable support among his people. So, the Pentagon's failure to send enough troops to take Baghdad fairly quickly could complicate the postwar occupation, to say nothing of the war itself."

While most observers were struck by the remarkable pinpoint accuracy of U.S. bombs and the care the U.S. military took to avoid civilian casualties, American left-wingers—along with most Arabic news channels—focused on what casualties there were, thus exaggerating their prevalence. Robert Sheer sneered in the *Los Angeles Times,* "We are told endlessly by our government's public relations machine that the 'greatest care' is being taken to prevent civilian deaths, as if good intentions matter to the child whose mother is killed."

The conflict did not last long enough to create a large body of antiwar literature. But the carping and nay-saying for those three weeks was shrill and insistent. Only when Baghdad had fallen did the liberals turn from what a terrible job the military was doing fighting the war to what a mess they were supposedly making of the occupation. But there were one or two who openly confessed their feelings. Blogger Andrew Sullivan quoted Gary Kayima, of the liberal website Salon.com, who wrote:

> I have a confession. I have at times, as the war has unfolded, secretly wished for things to go wrong; wished for the Iraqis to be more nationalistic; to resist longer. Wished for the Arab world to rise up in rage. Wished for all the things we feared would happen. I'm not alone. A number of serious, intelligent, morally sensitive people who oppose the war have told me they had identical feelings.

Very few were honest enough to acknowledge such feelings publicly, but Kayima is surely right that many in his profession shared them. Some of that negative energy could be felt in the reporting about the victory. Most networks and newspapers devoted one day's coverage to the fall of Baghdad and Iraqis celebrating in the streets. They then pivoted immediately to cover the supposed scandal of the looting of Iraq's national museum. The original stories suggested that 170,000 artifacts were stolen in a matter of days. Liberals were indignant. Why weren't the U.S. Marines assigned to protect the museum? Everyone knows that museums are targets of looting after a war.

The *Washington Post* gave prominent play to the resignations of two "cultural advisors" in the Bush administration. These were actually Clinton holdovers on a previously obscure panel called the President's Advisory Committee on Cultural Property. They issued a sarcastic statement saying, "While our military forces have displayed extraordinary precision and restraint in deploying arms—and apparently in securing the Oil Ministry and oil fields—they have been nothing short of impotent in failing to attend to the protection of [Iraq's] cultural heritage." The *Boston Globe* editorialized that "the awful truth is that the U.S. government bears a shameful responsibility for not preventing this crime against history." Eleanor Clift of *Newsweek* wondered aloud whether the United States might face war-crimes charges.

As the weeks passed, it became clear that the press had been duped by Iraqis eager to score points against the Americans. The "looting" turned out to be largely fictitious—most of the museum's artifacts having been salted away long before hostilities commenced. By early May 2003, estimates of the number of missing treasures had been revised from 170,000 to twenty-five.

The pattern persists. Make an accusation against the United States and you may confidently expect a respectful hearing among liberals. Attack the United States and some liberals will ask themselves what America has done to deserve it. Topple a vicious tyrant and many liberals will focus on your failure to protect a museum, or create a furor over the use of one questionable fact in the State of the Union Address. It is a perversity of our times that those who claim to be most dedicated to liberty are so alienated from the nation that best embodies it.

# None Dare Call It Victory

WHO WON THE COLD WAR? IT SEEMS absurd to pose the question, and yet, the past decade has become so clouded by revisionism and retroactive self-justification that a measure of clarity on the matter has been lost.

Some on the Left are now attempting to obscure the history of the period—to say that all of us were united in our opposition to Communism. That is false. It is so obviously false that it seems unbelievable that this theory has slid so smoothly onto the history shelves.

The West won the Cold War. The free nations defeated the totalitarian ones. The capitalists outperformed the statists. The believers outlasted the atheists. The United States of America, flawed and divided as we were, persevered to see the Union of Soviet Socialist Republics and nearly all of its satellites implode. For most Americans this was a welcome and even prayed-for conclusion to the Cold War. But for many others, it was neither hoped for nor celebrated.

## THE WALL IS TORN DOWN

More than a decade later it all seems a blur of images: joyful Berliners—no longer East and West—hacking away at the hated wall; statues of Lenin enmeshed in ropes toppling and crashing to the ground; Boris Yeltsin leaping onto a tank to the grateful cheers of ordinary Russians; and the hammer and sickle flag being lowered for the last time over the Kremlin. The world stood stunned as the Communist empire ended "not with a bang but a whimper."[1]

With dizzying suddenness, the great drama of the second half of the twentieth century, the great cause that cost dearly in life and treasure, and the great divider of American politics was...gone. The Cold War, which began in the weary years after the Second World War, when, in the words of Winston Churchill, "the whole world [was] divided intellectually and to a large extent geographically between the creeds of Communist discipline and individual freedom,"[2] was over.

The end was so abrupt that it gave rise to a kind of vertigo in the West. Officials in George Herbert Walker Bush's administration, so conditioned—as everyone was—to a world containing a terrible adversary, squinted like a person who has just emerged from a darkened theater into bright sunshine. Though the most dramatic and worldshaking events unfolded in Europe on his watch, the elder President Bush never offered public words to mark the momentous and longed for demise of the Communist colossus.

Certainly a degree of eye rubbing and incredulity was understandable. Almost no one—neither the most ferocious anticommunist nor the most appeasing fellow traveler—imagined that the Soviet Union and its subalterns would simply fold up their tents and declare bankruptcy. But that is exactly what happened.

The door had been left ajar by Mikhail Gorbachev, the Soviet leader whose ascension to power had caused such swoons in the West.

It was Gorbachev's hope to reform Communism and thereby save it, not to slay the monster from within. Still, there is no doubt that the liberalization he permitted gave heart to the oppressed peoples of Eastern Europe and reignited hopes for freedom that had not been tested for two decades. (Gorbachev did not take the same benevolent approach to the Baltic states two years later.)

As they had done in East Berlin in 1953, in Budapest in 1956, and in Prague in 1968, the people of Eastern Europe saw the light glinting through the unlocked door and rushed to pry it open further. They held candlelight vigils in churches and poured into the streets demanding free elections and other liberties. But this time the Soviet tanks did not roll in to crush them. Frantically plugging fingers into a thousand leaks in the dikes of repression that had characterized Communist rule, the puppet governments of Eastern Europe attempted over the course of the last months of 1989 to stanch the flow with concessions. In Poland, the government restored to the Catholic Church property that had been confiscated decades earlier. The Church was permitted to reopen Catholic schools. First Hungary and then Czechoslovakia agreed to multiparty elections (in other words, real elections). Vaclav Havel was released from prison. The regime in Budapest began to dismantle the 218-mile security fence that had kept Hungarians and other Eastern Europeans from escaping to Austria.

But as the Soviets themselves would discover in the following two years, freedom cannot be ladled with a thimble. Sensing that genuine liberty was within reach, the peoples of Eastern Europe, thwarted for forty years, rushed to embrace it—all of it. In three months' time, 120,000 East Germans had made their way to West Germany through Czechoslovakia, Hungary, and then Austria. As recently as February 5, 1989, East German border guards had shot and killed a man attempting to escape to the West.[3] Such barbarities had been common for more than forty years. But in the autumn of 1989, for the first time since the Iron Curtain locked up Eastern Europe tight, the Communist governments in those countries made no effort to prevent people from fleeing to the free nations of Europe. On the anniversary of the Soviet invasion of Czechoslovakia, thousands thronged Wenceslas Square. In the East German city of Leipzig, weekly prodemocracy demonstrations were held throughout the autumn.

Within months, the Communist regimes of Poland, East Germany, Czechoslovakia, Hungary, Yugoslavia, and Rumania dissolved. Bulgaria fell a year later. They fell, ironically, like the dominoes the West had once feared in Southeast Asia.

## BEWILDERMENT

The joy in Europe was incalculable. And though Americans too were swept up in the celebratory mood (even the left-leaning Leonard Bernstein traveled to Berlin to conduct a jubilant Beethoven's Ninth at the Brandenburg Gate), the American reaction was mixed. Though victory in the Cold War represented a triumph over a regime comparable to the Nazis in oppression, death, and destruction, there were no throngs of people in the streets, no ticker tape parades, and no generalized sense of triumph. Most Americans had believed that strife, or at the very least, prolonged tension with the Soviet Union, was an immutable fact of life. Some believed that the Cold War would inevitably lead to nuclear Armageddon. How then to account for the muted response to its peaceful conclusion?

In part, the United States had become preoccupied with other irritants—Panama's Manuel Noriega and, later, Iraq's Saddam Hussein attracted some of the finite attention Americans pay to events overseas. But the fall of the Berlin Wall was such an epochal event that the existence of distractions is inadequate to explain the neglect of it.

## REWRITING HISTORY

To understand the American response, one must look elsewhere. In the initial months after the Soviet Empire fell away, sheer vertigo gripped the American administration.

So unready were Secretary of State James Baker and President George Bush for a world without the Soviet Union that they responded to the initial liberation of the Soviet satellites as if our world, not theirs, was in danger of dissolution. Having seen the rapid unraveling of the puppet states in Eastern Europe, the Bush administration, and most American observers, nevertheless continued to believe that the Soviet Union would continue. Though anticommunists inside and outside of government energetically urged the administration to support the forces of democracy and reform within the Soviet Union itself, the Bush administration, along with most opinion makers, was stuck on Gorbachev.

There is little question that as between Brezhnev and Gorbachev, or in truth, between any Communist leader in the world and Gorbachev, the latter was a huge improvement. But a starving man will willingly eat tree bark. That doesn't mean he sticks with that diet when he finds a banquet laid out for him. The Bush administration bolstered Gorbachev long after it became apparent that his reforms, *perestroika* and *glasnost*, rather than reforming Communism, were bringing it down altogether. Calling for order and "stability," the Bush administration actually found itself to the left of reformers within the Soviet Union. On a visit to Kiev, President Bush cautioned the excited Ukrainians, who were on the verge of declaring their independence, against "the hopeless course of isolation."[4] The Ukrainian assembly voted for independence anyway—as did the Baltic states, Belarus, and many more in a recapitulation of the Eastern European experience. Within twenty-four months after the Berlin Wall was hacked to pieces by elated Germans, the Soviet Union itself was no more.

But while the Cold War ended more abruptly than anyone expected, its demise was very like its life in that Americans were, and remain, deeply divided about its meaning. And here perhaps is the answer to the mystery of America's tepid response to what should have been a triumphant historical moment.

Liberal opinion makers in America were not overjoyed by the Cold War's close. Those who had never believed in fighting the Cold War at all were quick to deny that anyone had won. The themes they sounded were (not surprisingly) first expressed by Mikhail Gorbachev himself, who traveled, as a newly private citizen, to the site of Winston Churchill's legendary Iron Curtain speech—Fulton, Missouri. In 1946, Churchill had characteristically captured in a few short words the darkness that was then enveloping Eastern Europe. "A shadow has fallen upon the scenes so lately lighted by the Allied victory," he told his American audience, "from Stettin in the Baltic to Trieste in the Adriatic, an iron curtain has descended across the continent."[5]

Hoping to provide a coda, Gorbachev offered a reform Communist's view of the conflict that had rent the world for nearly half a century. It was a mistake, Gorbachev argued, to speak of "winners and losers" in the Cold War. Instead, there was a "shattering of the vicious circle into which we had driven ourselves."

This idea—that the conflict was a thing with a life of its own over which no one had true control—was quite familiar and widely accepted in the West. Gorbachev further argued that the cause of the Cold War was to be found in the West's misinterpretation of Stalin. Stalin, Gorbachev insisted, had neither the intent nor the capacity to expand communist hegemony beyond Eastern Europe.

Gorbachev's interpretation is tendentious, but understandable. He, after all, presided over the demise of a vast empire and had become a living fossil. It was largely due to him (and for this he deserves a guarded measure of credit) that the Soviet Union expired when it did. But a great many opinion leaders in the West, who had no need to justify the Soviet Union, enthusiastically agreed with Gorbachev's assessment and approved of President Bush's reticence to declare victory.

Before analyzing the eagerness of the liberals to agree with Gorbachev that "nobody won" the Cold War, it is necessary first to consider the limp response of the (at least nominally) conservative president. How could he fail to offer fitting rhetoric for that most remarkable moment in human history? It was not the sort of victory perhaps that called for brass bands and confetti; we were not literally at war and the enemy did not literally surrender. But during the course of forty years, Americans had lost more than 100,000 lives fighting two wars against communist adversaries; we had spent billions on the military and had lived with hair-trigger tension and doomsday scenarios. We suffered all of this because we were determined that the second of two grotesque totalitarian systems that stained the twentieth century would not prevail.

President Bush was a great believer in personal relationships among leaders. He had grown somewhat fond of Gorbachev, and felt strongly that it would have been bad form to, in his words, "gloat" that "we had won, and they had lost."[6]

Of course, gloating would have been both bad manners and the wrong sentiment at such a moment. The Cold War was not a football match. Its conclusion was not an occasion for sportsmanship—good or bad. Instead, it was a moment that ought to have been marked by prayerful celebration and thanksgiving. For millions of people worldwide (though over a billion in China, North Korea, and Cuba remain unfree), the end of Communism meant the end of profound physical and spiritual suffering. For millions more it meant release from the fear of nuclear war. For the entire world, it meant the possibility of peaceful progress on all fronts. It would have been possible to capture this in words, if Mr. Bush had simply tried.

But if President Bush was inhibited by a misplaced chivalry, American liberals seemed to have had quite other concerns. From their point of view, the Cold War had been at best a foolish confrontation caused by groundless suspicion and paranoia on both sides, and at worst a long running example of American imperialism and reaction. Most American liberals had long since ceased to believe that we were engaged in a "Cold War" over questions of liberty versus tyranny, or good versus evil, and had instead adopted the view that the Cold War was simply a kind of superarmed madness, in which two "scorpions in a bottle"[7] threatened to destroy each other (and all of humanity with them) for no discernible reason.

In the case of some opinion leaders in the U.S., it is not too much to say that the outcome of the Cold War proved a disappointment. Robert Heilbroner, a liberal academic whose economics book *The Worldly Philosophers* was required reading on many college campuses, lamented that "the collapse of the Soviet system, hailed as a victory for human freedom, was also a defeat for human aspirations."[8] E. L. Doctorow, author of *Ragtime* and other big sellers, described America's conduct of the Cold War as "an act of national self-mutilation."[9] And *New York Times* columnist Tom Wicker, observing the celebration of freedom in Czechoslovakia, was quick to caution that the United States was beset by crime, traffic accidents, drug abuse, vandalism, and other troubles. "Freedom," he wrote, "is...not a panacea; and that communism failed does not make the Western alternative perfect, or even satisfying for millions who live under it."[10]

Frances Fitzgerald, author of *Way Out There in the Blue* and *Fire in the Lake*, expressed the liberal view succinctly when she told an interviewer, "I don't know any American Soviet scholar who believes that the United States ended the Cold War."[11]

Those who spent their adult lifetimes denying that the Cold War was worth fighting, those who could never see the point in defense spending, and those who urged conciliation if not outright appeasement of the Soviet Union in every significant East-West confrontation for forty years—are now attempting to rewrite history. It's a little reminiscent of the Stalinist style of history, in which inconvenient or liquidated historical figures were simply airbrushed out of history textbooks. (The textbook trade may well have been the only thriving industry in Soviet Russia.)

This rewriting is taking two forms. The first argues that the USSR's demise proves that the nation Ronald Reagan labeled an "evil empire" was really nothing of the sort, and the Cold War was an unnecessary and potentially catastrophic mistake. As Strobe Talbott, later deputy secretary of state in the Clinton administration, summed up the conventional wisdom in *Time* magazine, "A new consensus is emerging, that the Soviet threat is not what it used to be. The real point, however, is that it never was. The doves in the Great Debate of the past forty years were right all along."[12]

Talbott is exactly wrong. It was the hawks, the Cold Warriors, who were proved by events to have been correct all along. If the Soviet Union was a paper tiger not worth containing, why then are liberals today attempting to wrap themselves in the mantle of "cold warrior"—a term once wielded by them only as an epithet?

A second sort of revisionism concerns the role played by liberals themselves. When the statues of Lenin and Marx came crashing down in Eastern Europe and the Russians and Eastern Europeans were at last permitted to speak freely, the climate of opinion about communism became suddenly and dramatically chillier than it had been for thirty years. Vaclav Havel was as responsible as anyone for the renewed realism about communism. "The communist type of totalitarian system," he said speaking to the U.S. Congress in 1990, "has left both our nations, Czechs and Slovaks—as it has all the nations of the Soviet Union and the other countries the Soviet Union subjugated in its time—a legacy of countless dead, an infinite spectrum of human suffering, profound economic decline, and above all enormous human humiliation. It has brought us horrors that fortunately you have not known."[13] Before 1989, such rhetoric from the lips of any American risked being labeled "McCarthyite."

But the fall of the Berlin Wall changed the mood. With survivors of Communism testifying to its miseries, it was suddenly less acceptable to suggest that the two systems had been merely "competitors," or that the U.S. itself was to blame for Cold War tensions.

Responding to this new infusion of reality into a realm that had been clouded by appeasement and self-delusion for so many years, liberals adjusted their rhetoric. Dropping talk of scorpions and competing empires, liberals suddenly began to assert that foreign policy had been a simple matter during the Cold War.

They were led by President Bill Clinton, an expert at historical revisionism of every variety. In his 1997 State of the Union address, President Clinton said, "One of the greatest sources of our strength throughout the Cold War was a bipartisan foreign policy. Because our future was at stake, politics stopped at the water's edge." Similarly, speaking at the Council on Foreign Relations, Clinton's national security adviser Sandy Berger reminisced about a time when "friends were friends and enemies were enemies." On the campaign trail in 2000, former senator Bill Bradley said, "Until the fall of the Berlin Wall, we were sure about one thing: We knew where we stood on foreign policy."[14]

Statements like these were designed to erase roughly thirty years (in some cases sixty years) of sympathetic treatment of communism on the part of the American Left. But the record of their actual positions on matters from the nature of the Soviet system to the need for defense spending, to aiding anticommunist guerrillas around the globe, is available. It reveals what those who lived through it recall—that the question of how or even whether to challenge the Communists bitterly divided America for at least thirty years. And liberals were, almost without exception, inclined to excuse, justify, or ignore the grave sins of our adversaries while always calling down the harshest possible judgment on America.

Lenin is widely credited with the prediction that liberals and other weak-minded souls in the West could be relied upon to be "useful idiots" as far as the Soviet Union was concerned. Though Lenin may never have actually uttered the phrase, it was consistent with his cynical style. And, as the following chapters demonstrate, liberals managed, time after time during the Cold War, to live down to this sour prediction.

# CHAPTER ONE

# The Brief Interlude of Unanimity on Communism

*It is better to be a live jackal than a dead lion—for jackals, not men. Men who have the moral courage to fight intelligently for freedom have the best prospects of avoiding the fate of both live jackals and dead lions. Survival is not the be-all and end-all of a life worthy of man.... Man's vocation should be the use of the arts of intelligence in behalf of human freedom.*                                 —SIDNEY HOOK

IN MARCH OF 1983, PRESIDENT Ronald Reagan, speaking to the National Association of Evangelicals, used words that would resonate throughout the remainder of his presidency and beyond. Speaking of the "arms race," he said, "I urge you to beware the temptation of pride—the temptation blithely to declare yourselves above it all and label both sides equally at fault, to ignore the facts of history and the aggressive impulses of an evil empire, to simply call the arms race a giant misunderstanding and thereby remove yourself from the struggle between right and wrong, good and evil."[1]

The use of the term "evil empire" provoked a fusillade of contempt from the American liberal Left. Many news reports characterized Reagan's language as "strident" (or found observers who would). The Associated Press quoted Lord Carrington, former British foreign secretary, as condemning "megaphone diplomacy," and calling for "dialogue, openness, sanity, and a nonideological approach to the dangerous business of international affairs."[2]

Henry Steele Commager, then a professor of history at Amherst, was quoted in the *Washington Post* a few days later, identified only as a "distinguished historian," not as what he was: a well-known liberal intellectual. He condemned Reagan's speech as "the worst presidential speech in American history, and I've read them all. No other presidential speech has ever so flagrantly allied the government with religion. It was a gross appeal to religious prejudice."[3]

*Time* magazine's Strobe Talbott, who would later serve as deputy secretary of state in the Clinton administration, made his disapproval clear. "When a chief of state talks that way, he roils Soviet insecurities."[4]

Hendrik Hertzberg, a former speechwriter for President Carter and later the editor of the *New Republic* magazine, was beside himself. "Reagan's speeches are much more ideological and attacking than any recent president's speeches," he told the *Washington Post*. "Something like the speech to the evangelicals is not presidential; it's not something a president should say. If the Russians are infinitely evil and we are infinitely good, then the logical first step is a nuclear first strike. Words like that frighten the American public and antagonize the Soviets. What good is that?"[5]

The notion that that harsh criticism of the Soviet Union had to be stifled because it would lead to nuclear war was rarely stated as bluntly as Hertzberg did, but it was widely believed on the Left, and resulted in a tendency—evident until the day the Soviet Union closed its doors forever—to excuse, airbrush, and distort the aggressive and despicable acts of that regime.

Mary McGrory, a columnist for the *Washington Post*, never quite got over her amazement at Reagan's obtuseness. Months later, writing on a related matter, she noted that "The president...embarrasses them [members of Congress] with his talk of the Soviets as the 'evil empire,' but they think he has convinced the country that the communists are worse than the weapons."[6]

It was obvious to liberals that the exact reverse was the case. It made some observers almost panicky to think that Reagan actually believed what he said. George W. Ball had been undersecretary of state in the Kennedy and Johnson administrations. He, too, referred to the "evil empire" speech as proof that Reagan was dangerous and simplistic on foreign policy. Writing of events in the Middle East, he condemned Reagan for "obsessive detestation of what you call the 'evil empire.' . . . Mr. President, you have set us on a dark and ominous course. For God's sake, let us refix our compass before it is too late."[7]

"Primitive, that is the only word for it," fumed Anthony Lewis, of the *New York Times*. "Believers, Mr. Reagan said, should avoid 'the temptation of pride'—calling both sides at fault in the arms race instead of putting the blame where it belonged: on the Russians. But there again he applied a black and white standard to something that is much more complex. One may regard the Soviet system as a vicious tyranny and still understand that it has not been solely responsible for the nuclear arms race. The terrible irony of that race is that the United States has led the way on virtually every major new development over the last thirty years, only to find itself met by the Soviet Union."[8]

Of course, even if Reagan had said that the Soviet Union was "infinitely evil," and we, "infinitely good," as Hertzberg recalled the speech (Reagan had not said that), it hardly follows logically that the "next step" would be a nuclear first strike. Hertzberg is a Harvard educated editor and certainly capable of understanding this, but fear distorted liberal thinking.

Sovietologist Seweryn Bialer of Columbia University disapproved of Reagan's rhetoric. Yet he provided evidence that it had hit home among the aged bosses of the Kremlin:

> President Reagan's rhetoric has badly shaken the self-esteem and patriotic pride of the Soviet political elites. The

administration's self-righteous moralistic tone, its reduction
of Soviet achievements to crimes by international outlaws
from an "evil empire"—such language stunned and humiliated the Soviet leaders . . . [who] believe that President Reagan is determined to deny the Soviet Union nothing less
than its legitimacy and status as a global power . . . status . . .
they thought had been conceded once and for all by Reagan's predecessors.[9]

In retrospect, it is clear that denying legitimacy to the Soviet Union
was a stroke of brilliance—a moral challenge that resonated from
Berlin to Vladivostok. The Soviet Union did not deserve legitimacy.
But at the time it was heresy. In 1983, Communism was regarded by
everyone to the left of center in American politics as either a fact of
life to be accepted or a genuine humanitarian impulse that had, perhaps, gone a little too far. Reagan's rhetoric profoundly disturbed the
status quo because it made the moral case for anticommunism,
whereas the conventional wisdom held that the "mature" and "realistic" approach to Communism was accommodation, not confrontation. Liberals thought they had convinced everyone who mattered
that anticommunism was the thing to be feared, not Communism
itself. And then along came Reagan suggesting that the Cold War was
really a matter of good versus evil. Liberals who sought to deny the
moral dimension of the conflict were blindsided by Reagan's assertive
anticommunism and responded with calumny and contempt.

By 1983, Communism ranked far below many other evils most
liberals could name. Communism was certainly not worse than nuclear
war—and this conviction would form the scaffolding of many a liberal
position, from opposition to the MX missile to support for the nuclear
freeze and unilateral disarmament. Fear of nuclear war would also form
the unstated subtext of many liberal responses to Cold War challenges
in other realms as well. The desire to soft-pedal criticism of the USSR
or China, for example, sprang in part from secret sympathy with the
collectivist states, but also from fear of provoking them to anger.

Communism was not nearly as evil, most liberals believed, as the false charge of being a Communist. And, though they would probably never admit this openly, the clear implication of liberal/left-wing solidarity behind Julius and Ethel Rosenberg and Alger Hiss was that it was more distasteful to make a true charge of communism than actually *to be* a secret Communist. In Anthony Lewis's fulmination against Reagan's evil empire speech, he paused over Reagan's kind words for Whittaker Chambers, noting with incredulity that "he cites Whittaker Chambers as a moral arbiter!"

It was not always this way.

## THE ORIGINS OF THE COLD WAR

When the Bolsheviks hijacked the Russian Revolution in 1917, most Americans responded with dismay. It was understood that while the Russian Revolution had been a mass uprising with multifarious leadership, the Bolsheviks were thugs who used force to silence and eliminate the more democratic elements in the coalition of which they were a part.

But by the 1930s, when the U.S. and other industrialized nations were enduring the Great Depression—a crisis that many believed marked the "death throes" of capitalism—sympathy for Communism and Soviet Russia expanded markedly. Leftist writers like Beatrice and Sydney Webb and *New York Times* reporter Walter Duranty offered rosy assessments of Communism under Lenin and Stalin. (Years later, it was revealed that Duranty was actually being blackmailed by the Soviets throughout his stint in Moscow.) Journalist and Lenin admirer Lincoln Steffens famously pronounced in 1921, "I have seen the future and it works."[10]

Membership in the Communist Party USA reached its peak in 1939, at 66,000 registered members.[11] William Z. Foster, the Communist candidate for president in 1932 earned 102,785 votes.[12] But just when Communism was enjoying its golden age in America, word of Stalin's purges began to filter out of the Soviet Union. Thousands of loyal Communists, many of whom had performed leading roles in the October Revolution, were paraded before rigged tribunals. The most prominent victims went through the degrading spectacle of "confessing" to a list of errors and crimes against the state. The prisoners were assured that their own lives, or those of their families, would be spared if they confessed to these trumped up charges. Stalin pocketed the confessions—and then had them shot anyway.

Along with the purge trials came word of widespread repression and slaughter dwarfing anything visited upon the Russians by the tsars. Knowledge of the terror-famine, which took the lives of some ten million people, did not become widespread until many years later (in part due to Walter Duranty's dishonest reporting). In 1939, when Stalin signed a nonaggression pact with Hitler, carving up Poland and giving Hitler a free hand to attack the West, most Americans became firm anticommunists.

By the early years following World War II, a rough consensus had jelled about what was then unselfconsciously labeled the "communist threat." Republicans and Democrats, liberals and conservatives alike believed that the United States must resist the expansion of Communist tyranny. Though many on the left continued to make excuses for Stalin—notably Lillian Hellman, Owen Lattimore, Corliss Lamont, and I. F. Stone—the lesson of the century, mainstream policymakers agreed, had been Munich. British Prime Minister Neville Chamberlain's naive 1938 pronouncement after meeting Hitler that he had achieved "peace in our time" was viewed, in light of the terrible war that followed, as the most foolish and indeed cowardly act of diplomacy imaginable. In the postwar period, leading Americans were certain that appeasement of an aggressor was foolish and immoral and they were determined to avoid that mistake with the Communists.

Though the United States and the Union of Soviet Socialist Republics had been wartime allies against Hitler, America had not forgotten the Molotov-Ribbentrop Pact of August 1939, which had made World War II possible. (When Ribbentrop traveled to Moscow to negotiate the deal, he reported feeling as comfortable with the Communists "as among my old Nazi friends.")[13] Nor did the conduct of the Communist Party USA and its fellow travelers evoke sympathy. Fully in lockstep with Kremlin policy, the American Communists had faithfully pushed the antifascist line until the Hitler-Stalin Pact. They then pivoted 180 degrees to denounce Britain's "imperialist war" only to turn about again after Hitler invaded the USSR in 1941. They were not a true American party, but marionettes, dancing to a tune called in Moscow.

During the Soviet occupation of Poland, 15,000 Polish officers surrendered to the Russians. Their fate was the subject of considerable dispute for forty years. In 1943, 4,400 officers, found with their hands tied with wire behind their backs and bullets in the backs of their heads, were discovered by the Nazis at Katyn Forest. The Soviets denied the massacre and blamed the Nazis (who were certainly capable of the atrocity).

With time, the Soviets' lie was uncovered. In 1990, just before the fall of the Soviet Union, two more mass graves of Polish officers were found, bringing the death toll to more than 15,000. When the Soviet archives were opened fifty years after the massacre, a document ordering the executions was found. It contained the signatures of every member of the Politburo.[14]

In 1941, the Nazis surprised the Soviets by turning on them and invading Russia. The peoples of Ukraine, Belarus, Moldova, and other Soviet captive nations at first welcomed the Germans as liberators. But the savagery of the Nazis soon persuaded them that it was better to stick with the devil they knew. Later when the Japanese attacked Pearl Harbor and brought America into the war, the U.S. and the USSR became temporary allies.

While the U.S. under FDR's leadership took this alliance to heart, warming to "Uncle Joe" Stalin and condemning anticommunist sentiments as unpatriotic, Stalin took a different view. Even before the war's conclusion, the Soviet Union had activated a huge network of spies among its allies, principally the U.S. and Great Britain, in preparation for the time when they would revert to being enemies.

After the war, Stalin invited the free Polish government in exile, whose members had been patiently awaiting the war's end in London, to visit Moscow. When they arrived, they were arrested. At Yalta, Stalin had promised Roosevelt and Churchill that Poland would have free elections. But at war's end, he rigged the election to ensure a Communist victory. Roosevelt went to Warm Springs, where he died in April 1945, surprised that his famous charm had failed to work on Uncle Joe.

Truman entered the presidency naive and somewhat ignorant. He had absorbed the soft Roosevelt administration view of the Soviets. Confiding in a Missouri friend toward the end of the war, Truman said, "You are needlessly worried about Russian communism. I am worried about British imperialism."[15] But he was a quick study and soon took the measure of Stalin, coming to a more realistic view of him than his predecessor had.

Soviet troops were dispersed throughout Eastern Europe. Unlike the Americans, who rapidly demobilized as soon as hostilities ceased, the Soviet troops remained in order to install friendly governments throughout the region. Reports reached the West of mass arrests and liquidations. One such was the Swede Raoul Wallenberg, a saint of the war years. He had left a comfortable life with a banking family in Stockholm to risk everything to save Jews, and succeeded in saving the lives of thousands. (It was he, not Oscar Schindler, who deserved a major movie about his life.) Wallenberg was asked to come to a meeting with the Soviet "liberators" of Hungary at war's end. He went—and was never seen again. The Soviets arrested him and he lived out the rest of his life, which ought to have been one graced by tributes and prizes, in the misery of the Soviet Gulag. Wallenberg thus became a bitter symbol of resistance to both tyrannies of the twentieth century. Neither Nazis nor Communists could abide him. Both knew he was their enemy.

Though revisionists of the Cold War would later argue that America provoked the conflict, it requires a tendentious reading of history to yield this conclusion. Not only did America rapidly demobilize (except, of course, for troops in Germany in keeping with the Four-Power Agreement), the U.S. also volunteered to submit its arsenal of nuclear weapons to an international body so that nuclear power would be forever dedicated to peaceful purposes only. With the Baruch-Lilienthal plan, the U.S. proposed to turn over its nuclear monopoly to an International Atomic Development Authority "to which should be entrusted all phases of the development and use of atomic energy."[16] Most nations of the world enthusiastically endorsed this idea. But Stalin, who had spies at Los Alamos and elsewhere who were stealing the secrets of the atomic bomb, used his UN veto to kill the idea. It is impossible to imagine a clearer sign of America's peaceful intentions than the Baruch-Lilienthal plan. Yet it was ignored at the time by those keen to represent the Soviet Union as the victim of Western "encirclement," and later by historians who preferred to see the two superpowers as equally guilty of fomenting distrust and enmity.

In the next years, the Soviets renounced their nonagression treaty with Turkey and proceeded to make territorial demands on her, instigated a civil war in Greece, ordered the small Czechoslovakian Communist Party to seize power, and attempted to choke off West Berlin. President Truman responded to these provocations with the Anglo-American airlift, the Marshall Plan, and the North Atlantic Treaty Organization. Under the Democrat Truman, and with the support of Republicans in Congress, the U.S. went to war for South Korea, losing close to fifty thousand men.

Still, the domestic consensus that the U.S. ought to contain the Soviet Union held throughout the decade of the 1950s. It held despite the "excesses" of Joseph McCarthy, the inflammatory senator from Wisconsin who focused on the domestic danger of subversion and spying. The danger was real. As we now know from both Soviet and American sources, there did exist a high-level network of Communist spies in the American government, particularly during the 1930s and 1940s. Among these were Harry Dexter White, assistant secretary of the treasury; Alger Hiss, a high-ranking state department official; Donald Hiss, who held posts in both the treasury and state departments; and possibly Harry Hopkins, who was Franklin Roosevelt's most trusted advisor.

McCarthy was sometimes crude and ungentlemanly. He could also be sloppy and lazy. Whittaker Chambers, among others, kept his distance for that reason.

Though McCarthy flamed out and was censured by the Senate, liberals and leftists were able to use him for many years thereafter to discredit anticommunism of any stripe. A huge mythology about the "McCarthy era" has been spun suggesting that the senator plunged the nation into a reign of terror. The famous philosopher Bertrand Russell wrote in the *Manchester Guardian* (he was living in London in 1951) that the United States was a police state comparable to Nazi Germany or Stalin's Russia. "Nobody ventures to pass a political remark without first looking behind the door to make sure no one [is] listening. If by some misfortune you were to quote with approval some remark by Jefferson you would probably lose your job and perhaps find yourself behind bars...."[17]

Playwright Arthur Miller wrote *The Crucible* as an allegory, adopting the witch hunt theme that became the most popular shorthand for McCarthyism. It ran for 197 performances on Broadway. Woody Allen's movie *The Front* depicted the "Hollywood Ten" as innocent liberals smeared by hysterical anticommunists. Lillian Hellman, herself a longtime Stalinist who—among many other thoroughly contemptible acts—attacked the Dewey Commission for condemning Stalin's purges in the 1930s, wrote a widely acclaimed book in the 1970s called *Scoundrel Time*, which portrayed "McCarthyism" in lurid colors. William Phillips, editor of *Partisan Review* magazine, responded to Hellman in that magazine:

> Lillian Hellman's questions as to why we did not come to
> the defense of those who had been attacked by McCarthy

is not as simple as it appears. First of all, some were Communists and what one was asked to defend was their right to lie about it.... Another consideration was the feeling... that Communists did not have a divine right to a job in the government or in Hollywood.... Furthermore, it was not just a case of disagreeing with the Communists. They had branded us as the enemy. They were under orders not to speak to us. Their press called us every dirty name in and out of the political lexicon. And, of course, they were apologists for the arrest and torture of countless dissident writers in the Soviet Union and in other Communist countries.... How could Lillian Hellman not know these things?[18]

As Peter Collier and David Horowitz described it in 1989, "Thirty years after Joe McCarthy's death, 'McCarthyism' has become an ominous synonym for sinister authority and political repression.... Individuals and parties compete to brand each other with the scarlet M, using the term as the moral trump which automatically terminates arguments."[19]

There came a time when liberal publications simply declined to identify anyone as a Communist. When Angela Davis, the vice presidential candidate of the Communist Party USA spoke at Stanford University, the student newspaper identified her only as an "activist." When asked why they did not identify her with her own chosen party affiliation, the editor explained "That would be McCarthyism."[20] As late as 2001, a National Public Radio report on Arthur Miller would make a passing reference to "McCarthy's fascism" as casually as one would mention Cupid's arrow.

But the "McCarthy era" was a fable. Talk of an "era" is distortion, since it was not really an "era" at all. Only three-and-a-half years separated the Wheeling, West Virginia, speech that brought the senator to national prominence and the Army-McCarthy hearings that plunged him into disrepute. Further, the notion that McCarthy simply let fly with a spray of loose accusations was false. What he did was to cite actual government investigations of suspected Communists working in the government, particularly in the State Department, and demand that the administration do something about those who had been identified as possible security risks. There were scores of such people, and in each case, there were solid reasons to question their service in sensitive posts (though not to bandy their names about publicly).[21]

While McCarthy was certainly a bully, his influence (if not his fame) was limited. The lives of all but a tiny fraction of Americans were completely untouched by McCarthy or the "hysteria" that is said to have prevailed. Most Americans came to agree that McCarthy's style may have left something to be desired, but that the issue he helped bring to prominence—that of internal security—was real. Writing in 1996, liberal journalist Nicholas von Hoffman remarked: "...point by point Joe McCarthy got it all wrong and yet was still closer to the truth than those who ridiculed him."[22]

In his autobiography, philosopher and liberal Sydney Hook contrasted the McCarthy period with the war hysteria he recalled during World War I: "...Mob violence; abolition of due process; raids on private premises by public officials and vigilante groups; and arbitrary arrests.... The use of German became taboo, and instruction in it was dropped from the curriculum. Sauerkraut was renamed 'victory cabbage.' One became suspect if he had pictures of Bach or Beethoven on the walls, and Kaiser Wilhelm became the devil incarnate."[23]

Nothing of that sort shook America during the 1950s. The fact that "McCarthyism" is invoked so authoritatively is testimony only to the mythmaking power of the Left.

CHAPTER TWO

# The Consensus Unravels

*A great deal of intelligence can be invested in ignorance
when the need for illusion is great.*     —SAUL BELLOW

THE PRESIDENTIAL DEBATES OF 1960 are remembered now chiefly in aesthetic terms. Did Nixon's decision to forego TV makeup cost him the election? Did Kennedy's Addison's disease give him an unnatural but photogenic tan?

What is striking from a historical perspective is the degree to which the presidential race of 1960 centered on the Cold War. Both Kennedy, the Democrat, and Nixon, the Republican, were vigorous cold warriors, sparring about the islands of Quemoy and Matsu and other hot spots around the world. This was the last presidential race in which the Democrat would attempt to outdo the Republican in anticommunist zeal. By 1972, the Democratic Party would have abandoned the fight against Communism altogether.

In 1960 John F. Kennedy, determined to be just as hardline as Republican vice president Richard Nixon, went so far as to suggest that President Eisenhower had permitted a "missile gap" to grow between the U.S. and the USSR. (It turned out to be illusory.) Both men referred to Communist China as "Red China," a term that would, in just six or seven years, be considered a faux pas among enlightened people.

Once elected, President Kennedy issued the most comprehensive and bracing call to arms against Communism that had ever been uttered by an American president. In his inaugural address, he famously pledged that the U.S. would "pay any price, bear any burden, meet any hardship, support any friend, oppose any foe to ensure the survival and the success of liberty." This was the high-water mark of American anti-communism. And at the same time, it was the Kennedy administration that provided its low burlesque with half a dozen failed schemes to assassinate Fidel Castro, including exploding cigars.

Though revisionists from Ted Sorensen to Oliver Stone have attempted to suggest that President Kennedy never intended to get the U.S. involved in Vietnam, it seems clear from the historical record that he did get the U.S. in. (Stone engaged in wild allegations including the suggestion that the entire U.S. government conspired to kill President Kennedy.) Following a summit meeting with Khrushchev in which Kennedy had seemed weak, the president confided in columnist James Reston, "Now we have a problem in trying to make our power credible, and Vietnam looks like the place."[1] There were 16,000 American troops on the ground in South Vietnam when Kennedy was assassinated.[2] This is not to say that had he lived, he would have followed the escalation policy pursued by Lyndon Johnson, only that helping South Vietnam was completely consistent with Kennedy's foreign policy—to say nothing of his stirring inaugural. He took those fateful steps in Vietnam, which included aid, weapons, and training, in addition to personnel, with the full support and approval of the liberal establishment.

That establishment was still very much in Cold War mode. The *New York Times*, which would later condemn U.S. participation in the war as a crime and a tragedy, did not see it that way in 1961. A firm editorial spoke of "communist aggression" against South Vietnam that had been "launched as a calculated and deliberate operation by the Communist leaders of the North." The editorial continued: "The outlook in South Vietnam certainly gives no basis for optimism.... Free World forces, however, still have a chance in South Vietnam, and every effort should be made to save the situation. President Diem's call for additional financial help to build up his military forces, which has the support in Montana senator Mike Mansfield, of one of the best informed individuals in Congress on Vietnam affairs, should not go unheeded."[3] The very use of the term "free world forces" would soon be antique in liberal circles. Within a few years, liberals would refer to the "free world" only with irony and distance.

Though recommending financial help does not necessarily imply a willingness to declare war, it is worth noting that the *New York Times* watched silently as American participation on the ground increased dramatically. The *Times* did not protest when Attorney General Robert Kennedy, visiting Saigon in 1962, said, "We are going to win in Vietnam, and we will remain here until we do win."[4] The *Washington Post* warned North Vietnam in 1964 that "persistence in aggression is fruitless and possibly deadly."[5] As late as April 1965, the *New York Times* editorialized: "No one except a few pacifists here and the North Vietnamese and Chinese Communists are asking for a precipitate withdrawal. Virtually all Americans understand that we must stay in South Vietnam at least for the near future."[6] As Norman Podhoretz recounted in *Why We Were in Vietnam*, "the literary and intellectual community generally, which were later to become two of the great centers of antiwar sentiment...exhibited very little interest in Vietnam one way or the other."[7]

But as the war slogged on, this began to change. Under President Johnson, the U.S. pursued a hesitant and timid war policy. The U.S.

did not mine the harbor of Haiphong, nor hit key targets in North Vietnam. President Johnson picked the military targets himself—and his targets were chosen for their power to send messages, not for their ability to inhibit the North Vietnamese war machine. He hoped that gradual escalation would make the price of conquest too high for the Communists.[8]

As Peter Rodman records in *More Precious Than Peace*, the decision makers around Johnson were influenced by the recent experience of the Cuban Missile Crisis. In their reading, Kennedy's judicious use of "escalation" had ensured an American victory without firing a shot. Arthur Schlesinger Jr. rhapsodized about it in his book *A Thousand Days*: A "combination of toughness and restraint...so brilliantly controlled, so matchlessly calibrated, that dazzled the world."

Kennedy's foreign policy team became Johnson's foreign policy team, and they believed that in Vietnam, too, the "slow, judicious" use of power, "signaling clearly and cautiously [our] intentions" (the words are Robert McNamara's) would succeed. Bill Moyers, then a press aide to Lyndon Johnson, recalled:

> There was an unspoken assumption in Washington that a major war was something that could be avoided if we injected just a little power at a time. There was an assumption that the people in Hanoi would interpret the beginning of the bombing and the announcement of a major buildup as signals of resolve on our part which implied greater resistance to come if they did not change their plans.... There was a confidence—it was never bragged about, it was just there—a residue, perhaps, of the confrontation over the missiles in Cuba—that when the chips were really down, the other people would fold.[9]

In fact, the slow, judicious use of force, meant to signal our resolve, signaled instead the reverse—our reluctance to use greater force. As Rodman analyzed it, "If the hope was to *avoid* a major commitment of American troops and American power, as indeed it was, then the North Vietnamese were as capable of discerning this strategy as we were of devising it."[10]

The North Vietnamese endured the bombing and persevered, while at home, demoralization set in. It is difficult to recall a time when men have happily marched off to war in the name of achieving a truce. But as Podhoretz argues, Johnson was reluctant to ask Americans for sacrifice in Vietnam because his heart was in the Great Society. It was for his domestic "war" that Johnson used every bit of his persuasive power. He expended almost no effort in building support for the war being fought with bullets. And because Johnson was unwilling to fight to win in Vietnam, the war ground on and on, frustratingly inconclusive—but no less deadly for that. It may well be that sending American troops to Vietnam was a foolish policy—too great a commitment for the stakes involved. But it is certainly the case that Johnson failed to lay out the case for American intervention—and his negligence left the field open for those who had their own, quite unflattering interpretations of America's policy.

## THE ANTIWAR MOVEMENT

From the Left, a curdled view of the U.S. action in Vietnam bubbled up. For them, the war was proof that the United States itself was evil and corrupt. And while many on the left had always felt this way about America, their views found resonance for the first time with a more mainstream audience. From university educated (or partially educated) young people worried about their draft status (campus protests against the war, though supposedly based upon a profound sense of moral outrage, came to an abrupt halt when the draft was ended), to others who had no recent memories of Communist atrocities to gird them against leftist propaganda, many Americans paid heed. When the hippie generation gathered for its legendary rock concert at Woodstock in 1969, one group sang:

For it's one, two, three, what are we fightin' for?
Don't ask me, I don't give a damn—
The next stop is Vietnam.

The leftist critique of America's role in Vietnam set the tone for every Cold War debate that would follow for the next thirty-five years. It was during the Vietnam era that the consensus against Communist expansion was permanently shattered. It was during this time that American liberals—as distinguished from the hard Left—became not just neutral about Communist expansion, but contemptuous of cold warriors.

The Left's explanation of America's participation in the war was cartoonish and grotesque. It featured the U.S. in the role of international outlaw, interfering in a civil war for the sole purpose of propping up a corrupt dictatorship that would do its bidding. Some radicals went further, arguing that the U.S. had sent troops ten thousand miles across the globe to secure mineral rights in South Vietnam for large American corporations. Stokely Carmichael divined that the true reason for American intervention was "to serve the economic interests of American businessmen who are in Vietnam solely to exploit the tungsten, tin, and oil...."[11] They painted the North Vietnamese and Vietcong as our victims, and openly urged that the best possible outcome would be a Vietcong victory and a humiliating American defeat. Antiwar demonstrations did not chant merely "Bring the boys home," or "Make love, not war," but "Hey, hey, LBJ, How many kids did you kill today?" and "Ho, Ho, Ho Chi Minh."

Left out of this caricature, and increasingly left out of television reports from Vietnam, was any sense of the nature of the Communist enemy—an enemy that sometimes won the hearts and minds of the peasantry with stirring nationalist appeals, but also an enemy known to eviscerate those it suspected of betrayal. It was an enemy that used the civilian population for cover—thus purposely endangering the lives of innocent men, women, and children. There was never any acknowledgment among critics of the U.S. role that this tactic violated the Geneva Convention and was a deeply immoral way to wage war.

The profound tremor that went through American society starting in about 1965 was not just about the Vietnam War. Some deep wellsprings of dissatisfaction, petulance, and irritability were tapped by the war. All at once, everything about American society—from its "materialism" to its supposed "militarism"—was decried and despised. While there had always been critics of America, and while the brief popularity of Communism was understandable during the Great Depression, this shudder of revulsion for all aspects of American life came at a time of abundance and domestic progress. While leftist opinion had certainly held the United States in low regard for most of the century, in the mid-1960s, leftist anti-Americanism went mainstream. They condemned America for its poverty and at the same time for its consumerism. They demanded (and got) sexual license and grade inflation. They despised what they perceived as the drudgery of bourgeois life. Hillary Clinton, a Wellesley graduate in 1969, was typical of the young leftists of her day. In her much heralded graduation speech, she announced that the unimaginative, middle-class life her parents' generation had led was unacceptable. They would seek instead "more ecstatic" ways of experiencing the world.

The flavor of campus "protests" (many were really riots) was captured by Mark Rudd, leader of the "uprising" at Columbia University, in which radical students occupied five campus buildings for a week in 1968. Speaking to the university president, Rudd proclaimed, "You call for order and respect for authority; we call for justice, freedom, and socialism. . . . There is only one thing left to say. It may sound nihilistic to you, since it is the opening shot in a war of liberation. I'll use the words of [Black Panther] LeRoi Jones, whom I'm sure you don't like a lot: 'Up against the wall, m——f——. This is a stick up.'"[12]

Sympathy for Communism in the 1930s had many roots, not all of them disreputable. Some were attracted to Communism because its brutality gave them a thrill. But others joined the Communist Party out of the sincere, if misguided, hope that it would free mankind from age-old curses—poverty, ignorance, and want.

But sympathy for Communism in the 1960s was quite different. The New Left was aware of Communism's failures, its treachery, its tortures, and its murders. Their sympathy for Communism (they usually substituted the word "socialism") was an outgrowth of hatred for the United States. For the Left (and this came to encompass a significant part of the Democratic Party), the Vietnam War was immoral—probably the most immoral war in human history—and America as a nation was a depraved predator whose power represented a danger to the rest of the world.

The Vietnam War was actually not very different from the war in Korea that preceded it by a decade. Korea, too, was a civil war halfway around the world in which the United States had no direct interests at stake. Napalm was used. South Korea, like South Vietnam, was far from a model democracy. The Korean War involved severe civilian suffering, as most wars do. And yet there was no antiwar movement in the 1950s, and no doubting of the essential justice of the war effort. Why?

One reason is that President Eisenhower privately threatened the North Koreans with nuclear war when he negotiated an armistice, which meant that the war was over in three years.[13] (Later presidents would hesitate to use nuclear strong-arm tactics because the Soviet Union entered the nuclear club in 1949 and within a few years acquired a bristling arsenal.) Another is that memories of Soviet bullying over Berlin and Eastern Europe were still fresh in 1950. Stalin was still in control of the Soviet Union. And finally, most Americans still got their news from radio and newspapers in 1950. By 1965 television would bring Vietnam into the homes of Americans on a nightly basis.

The reporting that Americans saw on television about the Vietnam War dramatically affected public opinion. Coverage became progressively more tendentious and one-sided as the decade progressed. Following the famous Tet Offensive of 1968—which was presented as a military defeat, though in fact, it was a military victory—the American press became first cynical and then downright hostile about the war. Some reporters became understandably skeptical after the third or fourth government assurance that victory was around the corner. And it is true that neither Robert McNamara nor Lyndon Johnson was honest about his real doubts. In 1965 McNamara told Johnson that "victory" was far from sure.[14] Yet they continued to build up American forces. Nor did the military conduct itself intelligently. Commanders structured the fight primarily against the Vietcong, whereas hindsight makes clear that the real enemy was always North Vietnam. But the assumption that the government's lies were part of an overall corrupt venture in imperialism is hardly a logical inference. On the contrary, this accusation looks positively absurd in retrospect. If the United States truly were building an empire, for what earthly reason would it choose South Vietnam as the place for which to lay down lives and treasure?

## THE PRESS TAKES SIDES

That the commitment of half a million men to a war with no clear goal was an irresponsible use of government power is a fair criticism. And some critics of the war did frame the argument that way. There was much to criticize and even condemn about the conduct of the war. The army was not immune from the social breakdown then ravaging the civilian population. Discipline, good order, and morale suffered—particularly as the unpopularity of the war among opinion shapers at home filtered down to the soldier level in the jungles of Vietnam in a damaging feedback loop. Drug use was a serious problem, if not quite as endemic as it would later be portrayed.

But a significant portion of the American press corps in Vietnam reported the war as worse than a flawed campaign. They reported it as a disaster—and an American-made disaster at that. South Vietnam was not a struggling, democratic-leaning country fighting for its life;

it was a human rights nightmare. While it was true enough that America's ally was far from a perfect democracy, it was infinitely more liberal and humane than the Communist regime in the north, a comparison that was never included in such reports. Nightly packages of tape from Vietnam (they did not then have regular satellite transmission) featured the suffering of Vietnamese civilians, the tenacity of the Vietcong and North Vietnamese, American errors and folly, and the futility and absurdity of the American war effort. It was during the Tet offensive in 1968 that we first heard the immortal explanation by a soldier, "We had to destroy the village in order to save it."[15] It became, along with the photo of a desperate, naked child running down a road screaming after a napalm attack and the image of a South Vietnamese army officer executing a VC prisoner at point blank range, one of the endlessly repeated tropes of war coverage. Later, these images, along with the ignominious one of American helicopters lifting off the embassy roof in 1974 and the college coed at Kent State screaming over the body of a fallen student, became the stock footage of the Vietnam experience, defining it.

Each of the first three symbols, as B. G. Burkett records in *Stolen Valor,* was actually a misrepresentation. The photo of the child whose clothes had been burned off by napalm was reproduced worldwide as an example of the way the U.S. was using napalm against civilians. In fact, the photo dates from 1972, when nearly all American troops had left Vietnam. The child had been napalmed, but not by Americans. It was the South Vietnamese army that had called in air support after Communist troops had attacked her town. After the war, the government of communist Vietnam attempted to use Kim Phuc and her injuries for propaganda purposes. But Phuc declined. She later fled to the West and now lives in Toronto.

The photo of South Vietnamese general Nguyen Ngoc Loan executing the Vietcong prisoner made the rounds, and the history books, as evidence of the brutality of the South Vietnamese government America was defending. Photographer Eddie Adams of the Associated

Press won a Pulitzer Prize for the picture, and he knew the truth of the matter—namely that the Vietcong prisoner had only moments before murdered a Vietnamese policeman *along with his entire family*. Adams regretted that this photo became a rallying point for discrediting American participation in the war.

As for the famous "we had to destroy the village to save it" quote, it was reported by Peter Arnett, then a young reporter for the Associated Press, later a famed CNN correspondent reporting from Baghdad during the Gulf War. In 1968 Arnett reported a battle at a town called Ben Tre, where an American unit consisting of just twelve men with side arms was pinned down by enemy fire for two days and nights. Using howitzers and other munitions, some of which they had captured from the South Vietnamese, the Vietcong rained fire on the American position, killing many civilians and destroying large sections of the town, which was the provincial capital. Eventually, American gunboats were sent down river to relieve the trapped Americans. Several days after the town was secure, Peter Arnett was able to fly in on a military transport.

Arnett interviewed the two American officers present, army major Phil Cannella and air force major Chester L. Brown. Cannella, who later became disillusioned with all sides in the Vietnam conflict, was an American advisor to the South Vietnamese, and the ranking American officer on scene. He believes that he is the officer Arnett was quoting regarding the Ben Tre episode, but believes his comments were "taken out of context." He recalls telling Arnett that the Vietcong had destroyed the town and that it was a shame.[16] But Arnett's report made it seem that American forces had shelled the town and featured an *anonymous* American officer saying "We had to destroy the town in order to save it." An editor at the Associated Press seems to have substituted the word "village" for "town" to make it sound better, but Ben Tre was a town with multistoried buildings and paved streets.

In his memoir, Walter Cronkite, the "most trusted man in America," captured the press view of the war in a few short phrases:

Our TV cameras did record some—not all, but some—of the misery that the war brought to Vietnam. As I recall, we also reported some other disillusioning things about that war. . . .

A corrupt, incompetent, unpopular government that we were committed to support.

An allied army that often preferred not to fight.

A resourceful, dedicated enemy, resolved to struggle on regardless of casualties.

And the thoroughly reported lies and mistakes of our own leaders, whose political survival depended on making a war look good even as it turned bad.

The North Vietnamese, too, viewed these broadcasts, and understood the demoralizing effect they were having on the American public. Though Cronkite was widely regarded as occupying the center of the political spectrum, he always listed to port. In a Discovery Channel program entitled "Cronkite Remembers," the "most trusted man in America" had this to say about U.S.-Soviet relations: "I thought that we Americans overreacted to the Soviets. . . . Fear of the Soviet Union taking over the world just seemed as likely to me as invaders from Mars."[17]

B. G. Burkett was an army officer in Vietnam in 1968. He saw at first hand the media distortion of the war:

During an NVA [North Vietnamese Army] attack of a nearby village, several South Vietnamese civilians suffered burns when the enemy torched their homes. The injured civilians were brought to our dispensary for medical treatment. Two reporters appeared at the main gate to do a story on how the Americans "accidentally napalmed" the village. The village had not been napalmed. There had been no air strike. But the reporters had decided in advance what the story was. They wrote that the village had been napalmed.

Later, Burkett saw even worse examples:

In the middle of my tour, we went on alert when a unit of about four hundred NVA crossed a river north of us and were boxed in by American tanks and South Vietnamese infantry. A ferocious firefight erupted. The NVA refused to surrender; a dozen troops battled their way into the small village of Hoi Nai. The residents took refuge in the Catholic church, much more substantial than their tiny houses, which were made mostly of cinder blocks with tin roofs.

The American advisor to the South Vietnamese troops wanted our allies to take one or two platoons to rout the NVA stragglers, who had holed up in the abandoned houses. The South Vietnamese commander adamantly refused. That would put his troops at risk. He demanded that the Americans call in an air strike so he would not lose any men. The commander and the American advisor began arguing. Knowing that if anything in the hamlet was destroyed, we would rebuild it, the village chief weighed in, insisting on an air strike as well.

Meanwhile, a network television news correspondent and his cameraman roared up in a jeep. Based on the demands of the South Vietnamese, the American advisor called in an air strike. The scene was now incredibly dramatic: Fighters screaming in, dropping bombs, lots of noise and fires.

A few nights later, I saw the event on network television: the noise, the bombs, the dust rising, buildings burning. One of the cameras focused on an old Vietnamese lady, clearly terrified, huddling with a group on the front steps of the packed church, gripping a little girl, probably her granddaughter. The reporter's commentary was that Americans had destroyed a pro-American village, and he ended

by speculating, "Who knows how many civilians were
killed or wounded by American soldiers today?"[18]

In fact, Burkett explained, no civilians were killed that day, and
the U.S. army engineers were on the scene in the next few days
rebuilding the houses that had been destroyed and compensating the
villagers financially for their losses, "so much for a bicycle, so much
for a pig."[19] The villagers, all of whom were Catholic, had been
strongly pro-American before the fire fight and remained so after-
wards. Yet the television report left exactly the reverse impression,
and also planted the idea that many, perhaps hundreds of civilians had
been needlessly killed by American forces.

As columnist William F. Buckley Jr. pointed out at the time, more
South Vietnamese died attempting to save their nation from Communism than did Frenchmen resisting Hitler. Yet it became a common-
place of press coverage that the army of South Vietnam would not
fight, and that American conduct of the war was alienating the Vietnamese peasantry. A report on CBS's *60 Minutes* was typical. It
depicted a squad of marines torching a village with cigarette lighters
(a highly unusual U.S. action). Morley Safer offered this comment:

> The day's operation...wounded three women, killed one
> baby, wounded one marine, and netted these four prison-
> ers. Four old men who could not answer questions put to
> them in English. Four old men who had no idea what an
> I.D. card was. Today's operation is the frustration of Viet-
> nam in miniature. There is little doubt that American fire-
> power can win a military victory. But to a Vietnamese
> peasant whose home means a lifetime of backbreaking
> labor—it will take more than presidential promises to con-
> vince him that we are on his side.[20]

Hanson Baldwin, a former military writer for the *New York
Times*, wrote in 1971, "The extreme critics of the government have

lost their cool to such a degree that the Big Lie has become a part of our daily fare. The attempts to denigrate, to tear down, have one universal quality—to "poor-mouth" the United States, to attribute to our government, our military command, and our fighting men, a rapacity, cruelty, and ruthlessness that is a gross caricature of their true image."[21]

Did American soldiers commit atrocities in Vietnam? Without question the answer is yes. The whole world knows of My Lai, because an internal army investigation and prosecution brought it to light. But while much reporting and myth making since the war has given the impression that My Lai was typical (the philosopher Bertrand Russell accused the U.S. of "barbarous crimes committed daily"), there is good reason to doubt this.[22] In the first place, the soldiers who served in Vietnam have not reported any others. And unless you assume that all of the more than 2.6 million Americans who served in Vietnam were war criminals or covered up war crimes, a more reasonable supposition is that My Lai was a tragic aberration. Also, as Norman Podhoretz noted, "It is unlikely, given the number of antiwar journalists reporting on Vietnam, that if other such atrocities had occurred, they could have been kept secret."[23] A much worse massacre was committed by the North Vietnamese in the city of Hue. Between 4,500 and 5,500 civilians were murdered—many by being buried alive. But this atrocity received comparatively little attention.

Because war brings out the best as well as the worst in people, there were also inspiring moments of courage and tenderness. Burkett mentions Hugh Thompson Jr., an unsung hero. Thompson was a recon helicopter pilot who was flying over My Lai as the massacre was underway. He landed, and with his crew raced to the scene saving eleven Vietnamese civilians including a tiny baby and risking his own life. Burkett also quotes Morley Safer, who saw "a black sergeant, an enormous man, who had himself been raised in the most appalling conditions in Chicago, dancing across a dike during a mortar attack and pulling two children to safety."[24]

Misconduct by soldiers did not start in Vietnam. Nor did Vietnam stand out as a uniquely barbarous war in American history. As Burkett reports after examining the military records, between 1941 and 1946, in Europe alone, the U.S. Army (leaving aside the navy, army air corps, and all services in the Pacific theater) sentenced 443 men to death for murder, rape, and other capital crimes.

Nor were American soldiers in Vietnam particularly nasty to the Vietnamese peasants. The Green Berets alone dug 6,436 wells, repaired 1,210 miles of road, and built 508 hospitals and dispensaries.[25] American soldiers earned a reputation worldwide for friendliness and benevolence—even to defeated enemies. Certainly the prostrate Germans following World War II far preferred to be occupied by Americans than by Russians. The Japanese, too, discovered that a U.S. occupation, so feared in anticipation, was gentle beyond the dreams of benevolence (and to an enemy that had committed serious war crimes against our soldiers). Yet the Left—not least Hollywood—promoted the idea that suddenly, in Vietnam, American soldiers turned into crazed criminals.

Losing faith in a war whose reality they had themselves distorted, media big wigs followed up with bald antiwar propaganda. Following the Tet Offensive, Walter Cronkite prepared a broadcast on the conflict that was seen by millions. Standing on a huge map of Vietnam, Cronkite painted a bleak picture of the war effort, and concluded as follows:

> To say that we are closer to victory today is to believe, in the face of the evidence, the optimists who have been wrong in the past. To suggest that we are on the edge of defeat is to yield to unreasonable pessimism. To say that we are mired in a stalemate seems the only realistic, yet unsatisfactory conclusion....It is increasingly clear to this reporter that the only rational way out, then, will be to negotiate, not as victors, but as an honorable people who

lived up to their pledge to defend democracy, and did the best they could.[26]

According to Cronkite, President Lyndon Johnson watched this program and at its conclusion turned to aides and announced, "If I've lost Cronkite, I've lost middle America."

But actually, this was not true. While many Americans harbored doubts of various kinds about the war in Vietnam—particularly about the tentative and listless leadership demonstrated by Washington, majorities continued to believe that toughness was preferable to unilateral withdrawal. Johnson's popularity increased whenever he stepped up the bombing of the north.

Johnson misread the situation. Cronkite did not speak for middle America, but instead for the liberal intelligentsia and for a growing segment of the Democratic Party. Senator George McGovern, Democratic candidate for president in 1972 and a former bomber pilot from World War II, asserted that American bombing of North Vietnam was "the most murderous aerial bombardment in the history of the world."[27] A South Carolina delegate to the Democratic convention in 1972 asked George McGovern, "You want us to do all they [the North Vietnamese] demand and then beg them to give back our boys?" McGovern replied, "I'll accept that. Begging is better than bombing."[28]

## HO, HO, HO CHI MINH

McGovern's views would have made Jack Kennedy blanch, but he was following the lead of prominent intellectuals, like Susan Sontag, lionized in the *New York Times* as "a woman of letters."[29] A gifted writer, Sontag was among the first to succumb to the charms of the North Vietnamese. She traveled there during the war, one of a legion of "political pilgrims" (the phrase is Paul Hollander's) who journeyed to Communist countries in search of earthly paradise, and nearly always persuaded themselves that they'd found it. In Vietnam, Sontag was "struck by the grace, variety, and established identity of the Viet-namese.... Our impression of an effectively organized society, with its own genuine character, was deepened as we flew over the Chinese border into Vietnam. Below rolled miles of delicately manicured fields."[30] But Sontag was also explicit about why she traveled to North Vietnam: "Vietnam offered the key to a systematic criticism of America."[31]

"During the last years," Sontag wrote, "Vietnam has been stationed inside my consciousness as a quintessential image of the suffering and heroism of the 'weak.' But it was really America 'the strong' that obsessed me—the contours of American power, of American cruelty, of American self-righteousness."[32]

So overwhelmed was Sontag by the heroism and, as she saw it, "wholeness" of the Vietnamese, that she even thought she could see into their souls. "The phenomenon of existential agony, of alienation just don't appear among the Vietnamese."[33]

Ramsay Clark, former U.S. attorney general in the Johnson administration, traveled to North Vietnam and reported, "You can see no internal conflict in this country. I've seen none. You feel a unity in spirit. I doubt very seriously that I could walk in safety in Saigon or the cities and villages of South Vietnam, as I have here, because of the division and the confusion, and the lack of faith and belief there."[34]

The Reverend William Sloane Coffin Jr.—a cleric who would later head SANE/Freeze and a lifelong participant in left-wing causes—also traveled to Vietnam, where he experienced "a very special feeling for the North Vietnamese, a feeling I attributed to the fact that we were friends because we had deliberately refused to become enemies."[35]

Novelist Mary McCarthy, known for her asperity, showed little during her short visit (organized and initiated by the Communist government). She concluded that North Vietnam was infinitely preferable to South Vietnam. She never saw a child with a dirty face, she marveled, nor beggars, prostitutes, or squalor. "Wherever you go, you are met with smiles, cheers, hand clapping. Passers-by stop and wave to your car on the road."[36]

Though McCarthy had to acknowledge that North Vietnam had no free press, this did not dent her enthusiasm. She had ready excuses and explanations. Noting that the North Vietnamese regarded the right to criticize as just another "capitalist luxury," McCarthy agreed, saying, "The fact that you can read about, say, police brutality or industrial pollution in the *New York Times* or even in a local paper is nothing to be especially proud of, unless something concrete results, any more than the fact that you can read both sides about Vietnam and watch it on television."

Besides, it was America's fault that North Vietnam kept such a tight lid on information. "Until the Americans go home," she warned, there would be no respite "from the self-imposed rationing system in the realm of ideas that limited the [North Vietnamese] diet to what was strictly necessary to the national interest.... The Americans have blocked such possibilities for the young Vietnamese, and for the old too."[37]

Naturally, when the Americans fled the scene and the North Vietnamese were presumably freed from this "self-imposed rationing system in the realm of ideas," there was no sudden flowering of a free press. The Vietnamese government continued to be what it had always been—totalitarian. Miss McCarthy was not heard from on this subject.

Many "political pilgrims" to North Vietnam were enthralled by the gentleness they perceived in America's enemy. But not all. Frances Fitzgerald, author of *Fire in the Lake*, was impressed by the government's revolutionary ardor, even their incitement of hatred:

> In calling upon the peasants to hate their enemies, the Front cadres were asking them not merely to change their ideas but to disgorge all of the pent-up feelings they had so long held back, to fight what was to them the extension of

parental authority and stand up as equal members of the society.... The revolutionary project of the NLF [National Liberation Front]... was to use that released aggression as a creative force.[38]

War and terror were a form of therapy?

Jonathan Schell, a columnist for the *New Yorker* magazine, produced a relentless diatribe against the Vietnam War and the Nixon administration that continued for years. Denouncing what he called America's "rampage" in Vietnam, he scoffed at worries that we could "lose" Vietnam. "We cannot 'lose' it because we have never had it within our grasp, and because we have already done something much worse and much more decisive than losing it—we have destroyed it."[39]

Photos of Ho Chi Minh and Vietcong flags became fixtures at antiwar rallies during this period. For some Democrats (and they would later be called "conservative Democrats" for this stand) this was too much. Senator Henry "Scoop" Jackson recoiled, saying:

> I do not want to see the Democratic Party become a party which gives any aid and comfort whatever to people who applaud Vietcong victories or wave Vietcong flags. Our party has room for hawks and doves, but not for mockingbirds who chirp gleefully at those who are shooting at American boys.[40]

But Jackson's was a lonely voice in the Democratic Party. When President Nixon announced his decision to attack North Vietnamese sanctuaries in Cambodia, the nation was nearly paralyzed by protest. At Kent State University in Ohio, four students were killed when inexperienced National Guard troops, called in to maintain order, fired at the protesters. The image of a grief-stricken young woman, arm outstretched in a gesture of anguish over the body of a dead student, became another totemic image of the Vietnam era. In Washington,

D.C., between 75,000 and 100,000 marchers descended on the city to protest the "widening" of the war. A ring of sixty busses was required to encircle the White House and protect the president.[41] A group of government workers at the Peace Corps seized the building during this march, and flew the Vietcong flag from its roof.[42]

There was little acknowledgment in the mainstream American press, and even less among explicitly liberal writers and commentators, that while the U.S. had tried and convicted its own war criminals, the North Vietnamese had never done anything of the kind. Later, when American prisoners of war were released, Americans were shocked to hear that most had been badly mistreated while in captivity—starved, beaten, kept awake, denied necessary medical care, and kept in solitary confinement. The American people were stunned at least in part because the picture they had been offered of the North Vietnamese and their agents, the Vietcong, had been sanitized and romanticized.

When President Richard Nixon responded to a North Vietnamese offensive against South Vietnam in 1972 by mining the harbor of Haiphong, Democratic members of Congress were livid.

Congressman Ron Dellums, a California Democrat, took to the floor of the House:

> Those of us in Congress must have a great sense...of responsibility both to the Americans who are pawns in Mr. Nixon's games, and to the Vietnamese whose society we are turning into a smoking ruin....In the last year of the Second World War, after the Germans knew they were defeated, they went on an orgy of killing that exceeded the horrors of the earlier part of the war, haunting the conscience of mankind ever since. No longer able to impose our will in Southeast Asia, will our removal be in the same frenzied manner?[43]

Dellums was consistent with other left-wing critics of the war. The Nazi comparison was everywhere. Al Hubbard, a leader of Vietnam

Veterans Against the War, offered a "poem" in honor of his country, which, as was the fashion among the more rabid left-wingers, he spelled with a "k."

America
Now,
Before the napalm-scorched earth
Consumes the blood of would-be fathers
And have-been sons
Of
Daughters spread-eagled
And mothers on the run.
Reflect.
See what you've become,
Amerika.[44]

New York Democratic congressman Tom Downey was a strong proponent of abandoning South Vietnam: "The people of South Vietnam suffer as much from the cruelty, stupidity, and greed of their own government as they do from the war.... Our military aid is not saving these people; it is murdering them."[45]

Vietnam Veterans Against the War was extremely influential since it seemed to be a group of average veterans who had turned against the war because of the terrible things they had been forced to do or witness while wearing their country's uniform.

One highly influential publicity stunt was the "Winter Soldier Investigation." At a gathering in Detroit cosponsored by VVAW and a series of celebrities—Jane Fonda, Dick Gregory, Phil Ochs, Graham Nash, David Crosby, and Donald Sutherland—a procession of "veterans" came forward to tell tales of horror. One testified that he had been sent to "torture school" by the U.S. military. Another, that the entire war was "one huge atrocity" and a "racist plot." The testimony of those present was read into the Congressional Record by Senator Mark Hatfield, who asked the commandant of the U.S. Marine Corps to investigate.

When the Naval Investigative Service looked into these allegations, it discovered several interesting things. First, all of the "vets" who testified at the Winter Soldier event had been told not to cooperate with the investigation. One black marine who did cooperate was the one who had said the war was one "huge atrocity." He could not provide a single detail to investigators about any crimes. Further, he acknowledged that he had not even thought about atrocities until he left Vietnam. It turned out that a member of the Nation of Islam had significantly assisted his testimony.[46]

NIS investigators also found something else: Many of the veterans who were listed as attendees at the Winter Soldier meeting had not in fact been anywhere near the place. In other words, their identities had been pirated by the VVAW. Fake veterans thus testified to fake atrocities.[47]

VVAW did do some good for one real veteran, a decorated veteran at that. John Kerry, tall, handsome, and highly ambitious, was able to use VVAW as a launching pad for his political career. He had returned from service in Vietnam "as a rather normal vet," according to one friend: "He was glad to be out but not terribly uptight about the war."[48]

But the ambitious Kerry quickly gauged the political mood in Massachusetts and before long became the highly telegenic spokesman for VVAW. Kerry participated in one of VVAW's attention-grabbing gambits called Dewey Canyon III, a "limited incursion into the country of Congress." Members of VVAW marched on Washington wearing tattered fatigues and battle ribbons. After circling the Capitol and being turned away from Arlington National Cemetery, they held a candlelight vigil outside the White House. While someone played taps, the veterans stood up one by one and threw their medals over the fence—a defiant gesture of contempt for the war and for the nation that had asked them to fight it. Among those who threw his medals away was John Kerry.

Or did he? Years later, Massachusetts Democratic senator John Kerry changed his tune about Vietnam, assuring constituents that he was "proud" of his service. And his medals turned up, framed, on his office wall. A journalist with a longer than average memory questioned him about Dewey Canyon III, and the senator acknowledged that he had thrown someone else's medals over the fence. He was also cagey about what he meant by "proud," permitting listeners to assume he no longer believed the war was immoral. But on other occasions, Kerry has elaborated. After telling a *Newsweek* reporter that he was proud of his service, he added, "We should not be proud of what we did as a nation."[49]

For some opponents of the war, no amount of sympathy with America's Communist enemy was considered too much. Jane Fonda, daughter of privilege, traveled to Hanoi during the war. In obedience to the wishes of her North Vietnamese hosts, Fonda heaped scorn on the notion that American POWs were being mistreated. She posed gaily peering through the sight of an antiaircraft gun (presumably prepared to fire on American planes?), and made propaganda radio broadcasts. It is difficult to imagine what more she could have done to qualify as a traitor. Yet the nation that she and her husband at the time, Tom Hayden, daily accused of repression and criminality did not raise a finger against her. Quite the opposite. While always labeled "controversial," Fonda went on to enjoy a highly lucrative career as an actress and exercise guru. Hayden, whose rhetoric came in only one speed—overdrive—accused the United States of every crime in the book but never once voiced even the mildest criticism of any Communist country. "The real lesson of Vietnam," he said, " [is] what it may teach us about our genocidal history, about the real identity of American civilization as understood by its victims."[50]

Richard Barnet and Marcus Raskin, prominent antiwar activists and later officials of a prominent Washington, D.C., think tank, the Institute for Policy Studies, said, "The war itself was not an accident but the logical extension of a 'national security' policy of permanent war on a global scale. U.S. forces, agents, and spies operate in every country on the planet trying to make the world conform to an American vision of order that serves our pocketbook and our pride."[51]

Such suspicions about American intentions were not limited to extreme leftists. Harrison Salisbury, a *New York Times* columnist, traveled to North Vietnam in late 1966 and early 1967 and reported that the U.S. was deliberately targeting the civilian population. The effect of these reports, appearing in the nation's most prestigious and influential newspaper, cannot be overstated. Yet as Guenter Lewy reported in his book *America in Vietnam*:

> Only after the articles had appeared did a small number of persons learn that Salisbury, in effect, had given the authority of his byline to unverified Communist propaganda and that the *New York Times* had printed this information as though Salisbury had established it himself with his own on-the-scene reporting.... The dispatches dealing with the bombing of the city of Nam Dinh had borrowed extensively from a North Vietnamese propaganda pamphlet, *Report on U.S. War Crimes in Nam-Dinh City*....[52]

Noam Chomsky, a professor of linguistics at MIT who would later station himself on the farthest left edge of American political discourse, spoke for a large segment of the American academy when he said, "The Vietnam War is the most obscene example of a frightening phenomenon of contemporary history—the attempt by our country to impose its particular concept of order and stability throughout much of the world. By any objective standard, the United States has become the most aggressive power in the world, the greatest threat to peace, to national self-determination, and to international cooperation."[53]

When the war ended, the antiwar movement moved on to other things—unilateral disarmament, El Salvador, Nicaragua, the nuclear freeze—but few stopped to consider what happened to Vietnam and the region.

# The Communist Victory

Whether the Paris Peace Accords could ever have succeeded under any circumstances is impossible to gauge. As North Vietnamese negotiator Le Duc Tho shrewdly observed to Henry Kissinger during negotiations in Paris, "Before, there were over a million U.S. and puppet [South Vietnamese] troops, and you failed. How can you succeed when you let the puppet troops do all the fighting? Now, with only U.S. support, how can you win?"[54]

In fact, as events unfolded, it was even worse for South Vietnam than that. For while the Nixon administration had every intention of preserving South Vietnam's viability with military and other aid after an orderly American withdrawal of combat troops, Watergate soon robbed the administration of bargaining power with Congress.

Left-wing Democrats seized upon the Watergate scandal as a means to discredit Nixon and—though Nixon was not particularly conservative—American conservatism. With a president in disgrace, the antiwar Democrats in Congress could have their way. Starting with the 1974 budget, they refused to allocate another penny to Southeast Asia, and forbade U.S. military action "in or over" Indochina. So much for the threat of U.S. air strikes to punish North Vietnamese violations of the peace treaty. It wasn't enough that U.S. soldiers were out of it. The Democrats in Congress wanted the North Vietnamese to win. And they soon got their way.

The U.S. withdrew and then cut off all aid to its former ally. The South Vietnamese fought on, at first successfully rebuffing North Vietnamese attacks. But as David Frum recounts in *How We Got Here*,

> Congress chopped military aid to South Vietnam from $2.1 billion in fiscal 1973 to $1.1 billion in 1974 and $700 million in 1975—this while the price of petroleum was quadrupling. Commanders had to ration bullets. Artillery stopped firing. Trucks ceased moving. The South Vietnamese Air Force was grounded. When the Pentagon's

accountants tried to use a couple of hundred million dollars of unused appropriations left over from 1972 and 1973 to aid the South Vietnamese, Edward Kennedy mobilized a 43-38 Senate vote to forbid the expenditure. Giving South Vietnam the money it needed to survive, he said "would perpetuate involvement that should have ended long ago."[55]

When it was over, the Communists of the North were triumphant. Vietnam was at last "unified." And with the war's end came an end to many of the illusions propagated by the war's opponents. Stanley Karnow in his *Vietnam: A History* dismissed the notion that the Vietcong were anything but a tool of the North Vietnamese. He wrote: "By the fall of 1964, northern troops infiltrating into the south were enlarging Vietcong contingents.... Preparations to send North Vietnamese troops south had begun long before Lyndon Johnson seriously considered the introduction of American battalions.... The North Vietnamese were engaged in battle against Saigon government detachments months before the U.S. Marines splashed ashore at Danang."[56]

Whether critics of the war were truly deceived about the Vietcong or merely pretended to be is unclear. What is clear is that within days of their victory, the North Vietnamese swept aside their puppets. Vietcong leaders were ignored when it came to handing out powerful posts in the newly renamed Ho Chi Minh City—and those who protested found themselves in reeducation camps along with hundreds of thousands of others.

Estimates of the numbers of people packed off to such camps range between 200,000 and 1,000,000 out of a total population of twenty million.[57] Some were sentenced for only a few days, others spent years in prison. Reliable reports put the number of executions at 65,000. And this does not include those who died slowly in concentration camps.[58] The last prisoners did not emerge until 1986, eleven years after Saigon's fall. Conditions were atrocious. Prisoners had to subsist on a diet of only about 200 grams a day (7 ounces) of rice mixed with small stones.

Overcrowding was extreme. Seventy or eighty prisoners were squeezed into a cell built for twenty, and prisoners took turns breathing through one small air hole. In the tropical environment, the smells were ghastly, and skin diseases were rampant. Water was strictly rationed.

A plea from the "reeducated" made the rounds of Ho Chi Minh City. It was an oral testament, "signed" by forty-eight prisoners and then memorized and repeated.

> We,
> Workers, peasants, and proletarians,
> Believers, artists, writers, and patriotic intellectuals interned in different prisons across Vietnam,
> Wish first of all to express our debt of gratitude to:
> Progressive movements throughout the world,
> Workers' and intellectual struggle movements,
> Everyone who over the last ten years has supported the fight for human rights in Vietnam and supported the struggle for democracy and the freedom of oppressed and exploited Vietnamese citizens.
>
> The prison system of the old regime (which was widely condemned by international opinion) was quickly replaced by a more subtly planned system that is far harsher and crueler. All contact between prisoners and their families is forbidden, even by mail. The families of prisoners are kept in the dark about the fate of those in prison, which adds to the suffering and anguish. In the face of these humiliating, discriminatory procedures prisoners keep quiet, fearing that any objections they raise might result in further punishment for their relatives, who could be killed at any moment without their knowledge.
>
> Conditions inside the prisons are unimaginably bad. In the Chi Hoa prison, the official Saigon prison, 8,000 people under the old regime were kept in conditions that were

universally condemned. Today there are more than 40,000 people in the same prison. Prisoners often die from hunger, lack of air, or torture, or by their own hand....

There are two sorts of prisons in Vietnam: the official prisons and the concentration camps. The latter are far out in the jungle, and the prisoner is sentenced to a lifetime of forced labor. There are no trials, and hence, no possibility of using a legal mechanism for their defense....

If it really is the case that humanity at present is recoiling from the spread of Communism, and rejecting at last the claims of the North Vietnamese Communists that their defeat of American imperialism is proof of their invincibility, then we, the prisoners of Vietnam, ask the International Red Cross, humanitarian organizations throughout the world, and all men of goodwill to send us cyanide capsules as soon as possible so that we can put an end to our suffering ourselves. We want to die now! Help us to carry out this act, and help us kill ourselves as soon as possible. We would be eternally in your debt.

—Vietnam, August 1975–October 1977

However corrupt the South Vietnamese regime may have been, it did not cause hundreds of thousands of ordinary people to leave their homes, their loved ones, and their native country to take their chances on the open sea. Communist Vietnam did. By 1980, as many as 800,000 people—including a huge number of ethnic Chinese—had taken to small boats.[59] They were picked up by fisherman from many nations, as well as by American, British, Australian, and Israeli naval vessels, among others. But thousands were never rescued at sea and drowned. Some former apologists for the North Vietnamese became uncomfortable at these sights. But most did not. In 1983 Paul Cleary, head of the international office of the National Council of Churches, testified before Congress about Vietnam's reeducation camps. They resembled "a small tropical resort area," he said. "The entire process of reeducation is one reflecting the government's commitment to encouraging and enabling people to exercise their rights, restored as full participants in Vietnam's future."[60] Linda Ellerbee, commentator for CBS and ABC, joked about it when Hong Kong was about to be turned over to Communist China. "'These boat people,' says the government of Hong Kong, 'they all want to go to America.' Well, I swear I don't know why, do you? I mean, take Vietnam. Why would any Vietnamese come to America after what America did for Vietnam? Don't they remember My Lai, napalm, Sylvester Stallone? Clearly they have no more sense over there, than, say, Mexicans who keep trying to get into this country even though this country stole large parts of their country from them in the first place."[61]

But watching those desperate thousands bob on the open sea, a few of the vociferous critics of American participation who had been swept up in the wartime whitewash of the Communist side had second thoughts. Very few came to question their opposition to the war itself, but some rethought their indulgence of the Hanoi regime. Folk singer Joan Baez, who had been a fixture of many antiwar rallies (she was a Quaker and a pacifist), agitated to bring attention to the human rights nightmare that was unfolding in the former "victim of American aggression."

In 1979, Baez's organization, Humanitas, purchased a newspaper ad that ran in five large circulation dailies. An "Open Letter to the Socialist Republic of Vietnam" it read in part:

> Thousands of innocent Vietnamese, many of whose only "crimes" are those of conscience, are being arrested, detained and tortured in prisons and re-education camps....
>
> The jails are overflowing with thousands upon thousands of "detainees...."
>
> People disappear and never return....

People are used as human mine detectors, clearing live
mine fields with their hands and feet.

For many, life is hell and death is prayed for.

Though the ad was tough on Vietnam, it was also extremely dis-
paraging about the United States. "With tragic irony," these veterans
of antiwar rallies wrote, "the cruelty, violence and oppression practiced by foreign powers in your country for more than a century continue today under the present regime." They further wanted the
record to show that, "It was an abiding commitment to fundamental
principles of human dignity, freedom and self-determination that
motivated so many Americans to oppose the government of South
Vietnam and our country's participation in the war. It is that same
commitment that compels us to speak out against your brutal disre-
gard of human rights. As in the '60s, we raise our voices now so that
your people may live."

Perhaps she had to word it that way to coax any of her former
allies in the antiwar movement to sign. Still, the notion that cruelty
and oppression were foreign imports to Vietnam (by France and the
United States presumably) is a foolish and ignorant one. The Vietnamese, as even Frances Fitzgerald agrees in her antiwar book *Fire in
the Lake*, had an often savage history.

Baez had mailed the letter to 350 antiwar activists. Her yield of
signers was eighty-three. Among those who refused, many were livid
that Baez even raised the issue. "I had a lot of late-night phone calls,"
she recalled. They resented and suspected what they regarded as her
decision to "single out" Vietnam.[62]

It had been Ms. Baez's particular hope that Fonda would sign.
"Your name would mean more than any other," she told the actress
in a long letter. Fonda replied "Such rhetoric only aligns you with the
most narrow and negative elements in our country who continue to
believe that Communism is worse than death.... I worry about the
effects of what you are doing."[63]

Former presidential candidate George McGovern seemed to lean closer to Fonda than to Baez. Returning from a 1976 visit to Vietnam, he wrote in the *Nation* magazine: "[It] may not be what we would have chosen for the Vietnamese, but it was their country and choice, not ours. It may not be the best of all possible western worlds, but it is far better than our presidents warned it would be and infinitely better than endless war."[64]

But was it their choice? Without an endless stream of supplies and aid from the Soviet Union and China, the North Vietnamese would never have been able to dominate the South, as American liberals like McGovern were quite aware. Nor did Communist Vietnam ever hold a free election. The bald truth was that McGovern, Fonda, the Berrigan brothers, Pete Seeger, and even many more centrist Democrats affirmatively wished for a Communist victory in Southeast Asia—and have not to this day acknowledged the horror of what they achieved.

Liberals were horribly, catastrophically wrong about Vietnam. More importantly, they were wrong about the United States, and at the end of the day, it was the nature of their opposition to the war, not the opposition itself, that dishonored them. A reasoned and credible case could have been made, and here and there was, that sending ground troops to Vietnam was neither prudent nor necessary. But the shrill and vituperative attacks on a nation that, however awkwardly, was attempting to keep South Vietnam from the nightmare of Communist rule, were dishonorable and dishonest. And the willful blindness to the reality of the Communist enemy was a grave moral lapse.

CHAPTER THREE

# The
# Bloodbath

*To keep hope alive one must, in spite of all mistakes, horrors, and crimes, recognize the obvious superiority of the socialist camp.* —JEAN PAUL SARTRE

THERE WAS A SEQUEL TO THE VIETNAM War. For three years and eight months, beginning in April 1975, when the Communist Khmer Rouge movement defeated the forces of Marshall Lon Nol, the little nation of Cambodia was plunged into hell. Americans were not watching. They had turned their backs on Southeast Asia and were enjoying the first years of "peace." But for those left behind, for America's former allies, there was no peace.

The victorious Khmer Rouge (Red Cambodians) rolled into Phnom Penh and began a systematic war on the entire population so savage that it almost defies description. Estimates of the number of dead range between 1.5 and 2 million out of a nation of 7 million. At least one million were executed and another million died of starvation and disease that were the direct consequences of government policy.

In a nation of Cambodia's size, two million deaths represents between one quarter and one-third of the population. The Khmer Rouge caused this enormous number of deaths in less than four years. (Estimates are rough as Cambodia performed no census in the 1960s or 1970s.)

The Khmer Rouge cadres were very young, mostly teenagers, some as young as ten or twelve. Observers noted that some were barely taller than the guns they were carrying. Many had been kidnapped at the age of five or six from territories controlled by the Khmer Rouge and had known nothing but cruelty and indoctrination. They were country people, most of whom had never even laid eyes on a city. Sydney Schanberg, the *New York Times* correspondent who remained behind when the U.S. evacuated its embassy personnel from Phnom Penh, reported that Khmer Rouge soldiers played with the cars they captured like toys, switching the lights on and off for hours until the batteries died.[1]

Beginning on the day they took the capital, the Khmer Rouge undertook their massive and grotesque project of remaking society from scratch. They emptied the cities. Every single person in Phnom Penh including the lame, patients in hospitals on intravenous drips, elderly people with weak hearts and other ailments, mothers who had just given birth, pregnant women, and infants, was forced to leave. Some were told that the evacuation was a safety measure as the Americans were going to bomb the city. Others were informed that the evacuation was only for three days. Still others were told nothing but simply ordered by black-clad Khmer Rouge soldiers to take only what they could carry and march.

In the blazing heat, carrying a few pots and pans, some rice, whatever jewelry or cash they could hide (the cash quickly became worthless and was sometimes used as toilet tissue), two million people set out for the jungle. The smaller cities were emptied as well. Their destination unknown, all were forced to walk for days and days in the tropical heat. Those who had shoes counted themselves lucky. Stragglers and those who became ill were shot by the side of the road. The children cried with hunger and fatigue, but their parents could offer little comfort. If they were able, adults carried the little ones. If parents and children became separated, the adolescent soldiers showed no mercy. Family members were not permitted to search for one another but had to keep going. The bitter cries of lost children filled the air.

The indiscriminate murderousness of the regime was immediately apparent. John Barron and Anthony Paul reported for the *Reader's Digest* on the emptying of Phnom Penh:

Given the alternative of leaving or being shot, Dost Mohammad, an electronics salesman, departed on the morning of the 18th with his wife, six children, and his mother. They placed his mother, who was 75 and dying, on a bicycle which Dost and his son pushed. Some traffic was still moving on the street along which they walked, and a pedicab rolled past them. "Don't go on that side of the road!" a soldier shouted at the driver. The pedicab did not alter course, so a soldier killed the driver with machine gun fire. Another soldier riddled the driver of a Datsun who failed to heed, or perhaps even hear, an order to halt. A man driving a Chevrolet ignored one order but complied with a second command to stop. A young soldier thrust his rifle through the window of the car, then shot the driver through the heart, and he crumpled in the arms of his wife.... Racing to find his family before leaving the city, a man pedaled his bicycle into a street the communists had closed.... "I'm going to pick up my family," he yelled to guards. Without warning, a soldier sprayed him with machine gun bullets and he fell to the street dead.... Near the French embassy a French schoolteacher observed a communist patrol march from an alley through a line of refugees and by happenstance part a mother and father

from their children. The frantic parents protested and
sought to reclaim their children now on the other side of
the communist column. The patrol leader thereupon fired
a volley of rifle shots, killing both mother and father.[2]

Later, to save bullets, the Khmer Rouge switched to other methods of execution, including clubbing, asphyxiation, and, for Khmer
Rouge soldiers who had violated discipline, dousing the head with
gasoline and setting it afire.[3]

Loung Ung was five years old, one of seven children, when the
Khmer Rouge rolled into Phnom Penh. In her memoir, *First They
Killed My Father*, she recounts the march out of the city—her thirst,
her hunger, her blistered feet, her confusion that the soldiers had lied
when they said the family could return in three days.

Within a few weeks, mere blistered feet, sunburned skin, and
aching joints would have seemed like paradise to the Ung family, who,
like the rest of suffering Cambodia, were verging on starvation.
Shunted from village to village, forced to work punishing hours in the
tropical sun, and fed a thin soup containing only four teaspoons of
rice per day and an occasional piece of fish, they began to wither. All
around them, people were dropping dead. Often the bodies were simply left to rot. The stench was omnipresent. "To survive," wrote
Loung Ung, "my older siblings shake the trees at night, hoping to find
June bugs. The younger kids, because we are closer to the ground,
catch frogs and grasshoppers for food...."[4]

Overnight, the nation of Cambodia was transformed into one
huge slave labor camp. The Angkar ("Organization") was eager to
make a point about collective agriculture (like the Soviet Union and
China) and pushed its slave laborers harder and harder to produce
more rice for export. But if a starving person attempted to sneak a
handful of rice into his mouth, he would be shot. Even the food the
peasants grew themselves on their own land was off limits. Everything
belonged to the Angkar and eating a fruit from one's own tree became
a capital offense.

Pol Pot was quoted as boasting, "We don't have prisons, and we don't even use the word 'prison.' Bad elements in our society are simply given productive tasks to do."[5] In fact, there was an extensive penal system, but more than that, the entire country had been transformed into a penal system in which death was as casual as rainfall. There were countless capital offenses in democratic Kampuchea. Criticizing the regime in any way was enough to get one shot. Having any connection at all to the previous government or to any foreign nation would suffice. Failing to make abject confessions of error whenever corrected by a Khmer Rouge soldier was enough. Even submissively acknowledging, after promises of clemency, that one had, in an earlier life, supported other ideas, was sufficient to earn asphyxiation or decapitation. Of course, being caught in a lie was also a capital crime. Sexual relations outside of marriage could be a capital offense, depending upon the mood of the local Khmer Rouge (though soldiers routinely raped and killed young women). And, above all, stealing a morsel of food, particularly if it was not a first offense, was met with murder. This, in a nation the regime was purposely starving to death.

Loung Ung recounts with pain an episode early in her family's collective torture—before her family became separated—when they had managed to hide a stash of rice:

> Pa puts the rice in a bag, inside a container, and hides it beneath a small pile of clothes so that the other villagers cannot see it. On some nights when we really need it, Pa allows Ma to cook a tiny portion of the rice and mask the smell by burning damp, decayed leaves in the fire. This extra rice is our family's defense weapon against completely starving to death.
>
> One morning, Chou wakes all of us with her loud cries. "Pa, someone was in the container last night!" All eyes turn on the exposed rice container, the lid lies crooked on top and slightly ajar.

"Maybe some rats got into it and stole some. Don't worry, tonight I will seal it very tight," he says. "This rice belongs to all of us."

As Pa speaks, I know that he thinks someone in our family has stolen the rice. The story of the rat is not true and everyone knows it. Convinced that he realizes it was me, I hide my eyes from him. Shame burns my hand like a hot iron branding me for all to see: Pa's favorite child stole from the family. . . . Guilt weighs heavily on me. . . . I knew exactly what I was doing when I stole the handful of rice from my family. My hunger was so strong that I did not think of the consequences of my actions.[6]

She was six years old and her skin was hanging limp on her bones. Loung's father, a former official of the Lon Nol government, escaped the attention of the Angkar for one year. During that time, all nine members of the family became emaciated and exhausted by starvation. The Angkar moved everyone into the countryside without providing any sustenance or infrastructure. As the *Black Book of Communism* records, "Many were sent to unhealthy regions. . . . They had only the most rudimentary tools and were invariably given insufficient rations. They never had any technical assistance or practical training and were punished severely for failures of any sort, regardless of the reason. People with handicaps were simply treated as shirkers and executed."[7] Children did not attend school. As a Khmer Rouge official explained to Loung Ung, "Children in our society will not attend school just to have their brains cluttered with useless information. They will have sharp minds and fast bodies if we give them hard work. The Angkar cannot tolerate laziness. Hard work is good for everyone. Any kind of schooling carried out by anyone without the government's approval is strictly forbidden."[8] The Ung family fought starvation daily. They could think of nothing but food. Once, unseen by the Angkar soldiers or other villagers, they chanced upon two rabbits. It was the rainy season, and they had nothing dry with which to make a fire. And so they squirted some lime juice on the bloody meat and forced it down their throats. Loung nearly gagged, but a starving person will eat nearly anything, as she discovered in the following weeks when she and her siblings ate earthworms and a variety of insects. They saw their bodies bloat with edema, and their hair fall out in clumps. Keav, Loung's fourteen-year-old sister, was conscripted and taken to a work camp a few miles away. The work day was twelve hours long, but was often stretched beyond that to meet harvest goals. At those times, the black-clad Khmer Rouge would rouse workers at three or four in the morning and force them to work until eleven at night.[9] They were rarely permitted rest days, but when they were, these were taken up entirely by obligatory political lectures. Any sign of boredom or fatigue could lead to a beating or death. Keav succumbed to disease at the camp and was placed in one of the Angkar "hospitals." The *Black Book of Communism* records that, "Sick people were always suspected of malingering and were allowed to stop work only if they actually went to the hospital or the infirmary, where food rations were only half the normal size and the risk of epidemics was even higher. According to Henri Locard, 'the purpose of the hospitals was more to eliminate the population than to cure it.'" Without fresh water, adequate food, or medicine of any kind, stretched on the floor of a teeming "hospital" and deprived of her family, fourteen-year-old Keav died alone.

Within a few weeks, the Angkar found Loung's father. The soldiers told Ung to come out of his hut to help them. An ox had gotten stuck in the mud and they required his help, they said. Ung understood at once that he was facing his executioners. He asked leave to speak to his wife for a moment. Loung, who was six and did not understand what she was seeing, recorded his final moments:

> Pa and Ma go inside the hut. Moments later, Pa comes out
> alone. I hear Ma sobbing quietly. Opposite the soldiers, Pa

straightens his shoulders, and for the first time since the Khmer Rouge takeover, he stands tall.... I reach up my hand and lightly tug at his pant leg. I want to make him feel better about leaving us. Pa puts his hand on my head and tousles my hair. Suddenly he surprises me and picks me up off the ground. His arms tight around me, Pa holds me and kisses my hair. It has been a long time since he has held me this way. My feet dangling in the air, I squeeze my eyes shut and wrap my arms around his neck, not wanting to let go.

"My beautiful girl," he says to me as his lips quiver into a small smile. "I have to go away with these two men for a while."

"When will you be back, Pa?" I ask him.

"He will be back tomorrow morning," one of the soldiers replies for Pa. "Don't worry. He'll be back before you know it."

"Can I go with you, Pa? It's not too far. I can help you." I beg him to let me go with him.

"No, you cannot go with me. I have to go. You kids be good and take care of yourselves," and he puts me down.

Loong Ung never saw her father again. Later, her mother and baby sister were also executed.

The scale of devastation to the Cambodian people was of biblical proportions, but it was anything but a natural disaster. The lush climate of Cambodia had always permitted its population an adequate diet, and famine was unknown before 1975. The starvation Cambodia suffered under the Khmer Rouge was a political decision. It partook of some aspects of Nazi-style eugenics. They eliminated the aged, the infirm, the mentally handicapped, and those of non-Khmer ancestry. They reasoned that if people were weakened by starvation, they had less chance of mounting a challenge to the Communist Party. And people driven to desperation by hunger will quickly abandon all of the standards and virtues they had upheld in their former life. As Loung Ung recorded, "starvation had made us mad."

The Khmer Rouge were determined to create a new Khmer man just as the Russians had earlier built a "new Soviet man." To accomplish this, every vestige of the old civilization would be ruthlessly eliminated. Khieu Samphan, one of the Khmer Rouge leaders who had studied in Paris, wrote in his doctoral dissertation that the Cambodian economy and social structure would be renewed by tapping "the dormant energy in the peasant mass" against the cities.[10] The Khmer Communists were influenced by Stalin, Kim Il-Sung, and Mao. But they brought their own excruciatingly primitive stamp to the imposition of Communism in Cambodia.

Any sign of sophistication—knowledge of French (France had been the colonial power in the region for a hundred years), possession of eyeglasses, a typewriter, a watch, or a radio—could be a death warrant. Like the Nazis, the Khmer Rouge burned books, but they went further, smashing any artifact of twentieth-century life from telephones to phonographs to microscopes. Anything that could not be found in a typical peasant's dwelling was to be destroyed. Within a few months of their victory, the Khmer Rouge had eliminated all commerce, all transportation except movement of people by truck, all mail delivery (almost no one was in his original residence anyway), all education except mandatory indoctrination and "self-criticism" sessions, all religion, and all entertainment. Court records including real estate deeds, birth and death records, and corporate documents were all destroyed.

Family feeling and other human connections were also sternly discouraged. The Angkar forbade public displays of affection and discouraged acts of compassion and kindness. A man who attempted to keep his wounded son by his side in the fields was chastised by a Khmer Rouge soldier who told him, "You have individualist tendencies. You must shed these illusions."[11] When the same man attempted to visit his sick wife, he had to promise that the trip would not sap his energy, which properly belonged to the Angkar. When he tried to help a neighbor who was seriously ill and could not care for two small children, he was told, "You don't have a duty to help these people. On the contrary, that proves you still have pity and feelings of friendship. You must renounce such sentiments and wipe all such individualism from your mind. Go home."[12]

"We are making a unique revolution," one of Pol Pot's lieutenants exulted. "Is there any other country that would dare abolish money and markets the way we have? We are much better than the Chinese who look up to us. They are trying to imitate us, but they haven't managed it yet. We are a good model for the whole world."[13]

In 1975 there were more than five hundred doctors working in Cambodia (including Haing Ngor, who played Dith Pran in the movie *The Killing Fields*). By 1978 only forty remained alive.[14] By 1979, 42 percent of Cambodia's children had lost at least one parent (three times as many had lost fathers as mothers). Seven percent were orphans.

The carnage slowed when Vietnam invaded in 1979. Though the Khmer Rouge had been created by Hanoi as part of its effort to dominate all of Indochina, the two communist movements had become estranged. In 1979 the Hanoi regime invaded and installed its own puppet Communist regime to replace the Khmer Rouge. The new communist masters of Cambodia halted the wholesale slaughter but engaged in a systematic effort to starve the people in regions still controlled by the Khmer Rouge. Hanoi also kept at arms length international relief agencies attempting to feed and care for Cambodia's ravaged population. As a *Washington Post* editorial lamented, "Genocide, far from being an incidental product of a political decision almost seems to be *the* decision. Vietnam does not appear to want Cambodians to be saved from famine even in the parts of Cambodia it controls."[15]

# THE AMERICAN LEFT RESPONDS

Cambodia was not the first Communist state to massacre large numbers of its own people. That honor goes to the Soviet Union. It was the Soviet Union, the first Communist nation, that pioneered techniques like the terror-famine, mass deportations, concentration camps, and mass executions. Stalin had millions of his fellow citizens shot. Estimates of the total number of Soviet citizens murdered by the state range up to 20 million.[16] Under Mao Tse-Tung, an estimated 65 million Chinese were killed by execution, torture, and starvation. Vietnam is held responsible for one million deaths. North Korea is believed to have murdered 2 million.[17]

And so the Cambodian ordeal stands out only in proportional terms. The Khmer Rouge were not qualitatively different from other Communists, but they were more rushed. Communists have often been called "socialists in a hurry." The Khmer Rouge were Communists in a hurry.

It is therefore a mystery how liberal opinion makers could have so misjudged the nature of the Communist movements that stood to triumph if the United States withdrew all aid to noncommunist governments. But misjudge it they did. Even as the Khmer Rouge were shooting their way towards Phnom Penh, *New York Times* columnist Anthony Lewis wrote, "Some will find the whole bloodbath debate unreal. What future possibility could be more terrible than the reality of what is happening to Cambodia now?"[18] There is no doubt that after ten years of war, Cambodia's people had suffered. But Mr. Lewis had an impoverished imagination. So, too, did the *Washington Post*, which editorialized, "The threatened 'bloodbath' is less ominous than a continuation of the current bloodletting." The *Los Angeles Times* urged the cutoff of funds to the Lon Nol government "for the good of the suffering Cambodian people...."[19]

President Gerald Ford and Secretary of State Henry Kissinger urgently pleaded with Congress simply to permit the noncommunists in Indochina to defend themselves. The Democrats in Congress were adamant that this would not happen. Michigan Democratic congressman Bob Carr was typical:

> If we really want to help the *people* of Cambodia and the *people* of South Vietnam, is it not wiser to end the killing? Since most credited analysts of foreign policy admit that the Lon Nol regime cannot survive, won't the granting of further aid only prolong the fighting and, with it, the killing?[20]

Senator Chris Dodd, a Connecticut Democrat, sounded a similar note: "President Ford has tried to make this an issue of abandoning an ally.... The greatest gift our country can give to the Cambodian people is not guns but peace. And the best way to accomplish that goal is by ending military aid now."[21]

Congressman Tom Downey, a New York Democrat, was unmoved by talk of a bloodbath. "To warn of a new bloodbath is no justification for extending the current bloodbath."[22]

*New York Times* foreign correspondent Sydney Schanberg filed dispatches heaping scorn on the notion that a Communist victory was anything to dread. On April 13, 1975, just a week before the Lon Nol government fell, the *New York Times* ran a front-page story by Schanberg saying, "for the ordinary people of Indochina ... it is difficult to imagine how their lives could be anything but better with the Americans gone."[23]

As for the Khmer Rouge, Schanberg shared the then prevalent view that they were composed of a coalition of disparate elements, some communist and some not. (This theme was a constant liberal refrain throughout the Cold War.) Schanberg described Khieu Samphan as a Communist, but also as "a nationalist with a reputation for integrity, incorruptibility and concern for peasants, who is highly respected among non-Communist Cambodians." Schanberg was confident that there was nothing to fear from the Khmer Rouge. Their brutal behavior in the areas of Cambodia they controlled before 1975, was, he assured *Times* readers, "not…widespread" and would not be "need[ed]" once they controlled the entire nation.[24]

Even after the Khmer Rouge took the capital, Schanberg continued to ridicule the idea of a bloodbath:

> Another prediction made by the Americans was that the communists would carry out a bloodbath once they took over—massacring as many as 20,000 high officials and intellectuals. There have been unconfirmed reports of executions of senior military and civilian officials, and no one who witnessed the takeover doubts that top people of the old regime will be or have been punished and perhaps killed or that a large number of people will die of hardships on the march into the countryside. But none of this will apparently bear any resemblance to the mass executions that had been predicted by Westerners.[25]

That report was filed from Bangkok on May 8, three weeks after the Khmer Rouge takeover of Phnom Penh, an event Schanberg had witnessed with his own eyes. Even if Schanberg did not see any of the thousands of peremptory executions carried out in the opening days of Khmer Rouge rule, didn't the wholesale evacuation of the capital, a war crime according to the Geneva Convention, give him pause about the nature of the Khmer Rouge? Evidently not. For while Schanberg did quote one "Western observer" who watched the exodus from Phnom Penh from the French embassy and exclaimed, "They are crazy! This is pure and simple genocide. They will kill more people this way than if there had been hand-to-hand fighting in the city,"[26] he himself did not incline toward that view.

How can any reasonable person witness the wholesale evacuation of a city of two million people and not understand the enormity of what he is seeing? A city is an intricate network of millions of contacts between individuals and businesses. Banks, grocers, police stations, hospitals, farmers markets, gas stations, pharmacies—everything, in short, required for the functioning of an economy relies upon permanence and order. To empty a city—particularly by driving people out on the road in the blazing tropical heat—is an obvious death sentence to the very sick, the elderly, and the very young. But it is also a death sentence for many more—diabetics and other ill people who rely on a steady supply of medicine, those recuperating from illness who need rest and fluids to recover their health, pregnant and nursing mothers (and their babies) who cannot endure severe exertion or dehydration, and many more. Nor would city dwellers have been the only victims. What of farmers and fisherman deprived of a market in which to sell their goods?

Even if the Khmer Rouge had not gone on to commit genocide in other ways, the emptying of Cambodia's cities would stand as a crime against humanity. And certainly it was a clear signal of the regime's contempt for human rights.

But Sydney Schanberg, one of the few Americans to witness the Khmer Rouge takeover, was not thoroughly repelled by what he saw. Instead, as he explained to *Times* readers, "In almost every situation we encountered during the more than two weeks we were under communist control, there was a sense of split vision—whether to look at events through Western eyes or through what we thought might be Cambodian revolutionary eyes." Schanberg went on:

> Was this just cold brutality; a cruel and sadistic imposition of the law of the jungle in which only the fittest will survive? Or is it possible that, seen through the eyes of the peasant soldiers and revolutionaries, the forced evacuation of the cities is a harsh necessity? Perhaps they are convinced that there is no way to build a new society for the benefit of the ordinary man, hitherto exploited, without literally starting from the beginning; in such an unbending view people who represent the old ways and those considered weak or

unfit would be expendable and would be weeded out. Or
was the policy both cruel and ideological?[27]

Though Schanberg did not endorse the Khmer Rouge action, his
attempt at seeing it "through revolutionary eyes" would be inconceivable in a different context. If Schanberg had been a foreign correspondent in Berlin in 1938, would he have attempted to see
*Kristallnacht* "through National Socialist eyes"? Clearly not. In fact,
Schanberg would have despised anyone who wrote such a thing. Yet
for movements calling themselves "revolutionary," any crime—no
matter how repulsive—could at least get the benefit of an open mind
from the correspondent from the *New York Times*. Mr. Schanberg
was also rewarded with a Pulitzer Prize for his Cambodia reporting,
and became a hero of the movie *The Killing Fields*.

## UGLY AMERICA

Later, when the full story of the Communist nightmare endured by
Cambodia had become common knowledge, those who had appallingly
misjudged the Communists found a way to shift blame for the crimes
away from the perpetrators. They blamed the United States.

One of the first to adumbrate this soon-to-be widespread interpretation of events was U.S. senator Claiborne Pell, a Rhode Island
Democrat. Pell cast doubt on the stories of North Vietnamese atrocities at Hue:

> There is some question as to how many people would actu-
> ally suffer if South Vietnam came under Communist
> administration. The example is often cited of Hue, which
> after having been occupied by the North Vietnamese was
> found to have mass graves with many bodies in them. We
> had assumed these people had been shot by the North
> Vietnamese, although a story has surfaced to the effect that
> our bombing and razing of a great portion of Hue resulted

in many deaths and that these victims too were buried in
the same open trenches.[28]

This view was not then supported by any facts nor have the inter-
vening years provided any. It was the fantasy of a mind corrupted by
the idea that America—or more specifically, the administration of a
Republican president—was responsible for every evil in the world.

Certainly Anthony Lewis belongs among the all-stars of the group
Jeane Kirkpatrick would later dub the "blame America first" school.
A passionate writer, and to judge by his choice of topics, a man of
some sympathy for his fellow creatures, Lewis could see perfidy only
in America and her allies. On April 21, 1975, when the Khmer Rouge
were marching millions of people into the jungle, Anthony Lewis used
his *New York Times* column to fulminate against President Ford for
having said "I wish to express my admiration for the Cambodian government [Lon Nol] leaders." Ford also spoke of "compassion" for
Cambodia's people:

> Admiration! For politicians who fattened on American aid
> as their people starved. For military officers who made
> their soldiers pay for the rice sent by the United States—or
> supplied no food, so that some were reduced to cannibal-
> ism. For officials who forced United Nations relief agencies
> to pay $100,000 for the privilege of flying in powdered
> milk for starving Cambodian children. For a government
> that, even as it fell, was arranging to have a New York
> bank pay $1 million to Lon Nol out of a Cambodian
> account? Compassion! The word should burn in the
> mouths of American leaders after what we did to Cambo-
> dia. We dragged a peaceful country into a useless, devas-
> tating war—for our own purposes.[29]

It is certainly the case that the Lon Nol government was corrupt.
And it would be unfair, when hindsight shows the true misery into

which Cambodia was then falling, to chastise Lewis for not having fully foreseen it. But Lewis's default setting, in a long and influential career, has always been acid condemnation of the U.S. and her friends, and inattention or indifference to the shortcomings of our foes. Not only does this tilt his analysis; it undermines it. If Lewis were not so single-mindedly focused on America's sins (which obviously do exist), he might have taken the time to study the behavior of Cambodia's role models. If he had, he might have reflected on the horrors practiced by the North Koreans and the Chinese Communists, the two regimes for whom the Khmer Rouge professed the most admiration. And he might then have considered that if Cambodian officers were forcing the UN to pay bribes in order to import milk for babies, *at least the babies were getting the milk*. And even if it were true that Lon Nol was salting away some aid money for his own use—such petty corruption was far less devastating to the average Cambodian than the totalitarian nightmare of Khmer Rouge-style Communism.

Americans know of Cambodia's ordeal chiefly through the award-winning film, *The Killing Fields*. David Puttnam, the director, saw his movie as a cry against "extremism" of all forms. But in effect, if not in intention, the deeply affecting movie was an exercise in moral equivalence.

Though it accurately portrayed the depredations of the Khmer Rouge in the second half, the first part of the movie was an explicit rendering of the "blame America" viewpoint. Without explicitly saying so, the film presented the theory offered by journalist William Shawcross in his 1979 book *Side-show*.

Shawcross did not deny the Cambodian holocaust, as some on the left did. Richard Dudman, for example, who reviewed the book for the *New York Times Book Review* was critical of Shawcross for being too hard on the Khmer Rouge! He wrote: "Unfortunately, Mr. Shawcross accepts uncritically the testimony of Cambodian refugees about life...under the Pol Pot regime.... Two weeks of recent eyewitness observation inside Cambodia indicated that food, clothing and shelter were adequate under the strange regime and that working conditions, while hard, seemed by no means intolerable."[30]

Shawcross was not such a fool. But he did advance the argument that the United States was ultimately to blame for what befell Cambodia, because it was the U.S., he claimed, that had brought the peaceful land of Cambodia into the Indochina war in the first place. Then, having shattered Cambodia's neutrality, the United States commenced a bombing campaign that drove the Khmer Rouge mad and led to the disastrous policies of the mid-1970s. In *Side-show*, Shawcross writes:

> All wars are designed to arouse anger and almost all soldiers are taught to hate and to dehumanize their enemy. Veterans of the combat zone are often possessed of a mad rage to destroy, and to avenge their fallen comrades. It does not always happen, however, that victorious armies have endured such punishment as was inflicted upon the Khmer Rouge. Nor does it always happen that such an immature and tiny force comes to power after its country's social order has been obliterated, and the nation faces the danger of takeover by a former ally, its ancient enemy.[31]

In the movie, after agonizing footage of devastation caused by American bombs and the introduction of a particularly unlikable American officer (nowhere does the moviegoer learn that the war was introduced to Cambodia by anyone other than Americans), the viewer is treated to an almost verbatim version of the Shawcross thesis. Shredding papers in preparation to abandon the U.S. embassy in Phnom Penh, the deputy chief of mission is seething with anger. Through gritted teeth he tells Schanberg:

> What pisses me off is that this country has a lot of strengths and a lot of faults and we've done *nothing* but play to its faults. I'll be damned glad to get outta here. This thing had dragged on too long for it to end in all sweetness and light. And after what the Khmer Rouge have been

through, I don't think they're going to be exactly affectionate toward Westerners.[32]

The viewer is thereby instructed that the horror to come is traceable to the United States.

The notion that the U.S. was the first to violate Cambodia's neutrality was so widespread in the 1970s that it was simply taken as given. When David Frost interviewed Henry Kissinger on national television soon after the release of Shawcross's book, he began a question about Cambodia by saying, "We started to destroy a country." Kissinger attempted to correct the record—"We did not start to destroy a country..."—but Frost cut him off.[33] In 1979 *Washington Post* columnist Richard Cohen would casually refer to the U.S. "pounding [Cambodia] literally back to the Stone Age."[34]

It is a remarkable species of solipsism that can understand Cambodia's agony only as the result of American policy. Here was a textbook case of dedicated ideologues determined to remake their nation in a dystopian mold, determined to move further and faster toward "total communism" than their neighbors and rivals, the Vietnamese and Chinese—and the only fingerprints Western analysts can detect are those of Richard Nixon and Henry Kissinger! The charge is degrading and false.

It was North Vietnam who "widened" the war in Vietnam, not the United States. And it began when many of those who condemned Nixon as a war criminal were still supporting the war effort—before 1965. It was then that Hanoi created the Khmer Rouge and Pathet Lao with the ultimate aim of conquering all of Indochina for Communism (and, of course, for Vietnam). Shawcross and his acolytes had argued that the U.S. was to blame for the 1970 coup in which Prince Norodom Sihanouk was deposed. It was U.S. bombing of unpopulated regions in the east, they maintained, that brought about the coup.

Sihanouk's fall *was* a blow to Cambodia, but far from having been caused by American actions, it was the consequence of his own decisions. Sihanouk was constantly walking a tightrope as leader of a weak and vulnerable country. He would sometimes defy the North Vietnamese and cozy up to the West and at other times flatter the Communists and work that angle. In 1970, he made a decision to look the other way when the North Vietnamese invaded his country and took over several eastern provinces. The invasion was bitterly resented in Cambodia and the nation was gripped by anti-Vietnamese fever (the two countries were ancient enemies). Sihanouk attempted to balance his pro-Hanoi tilt with a nod to the Americans—telling Kissinger that he would not object to American bombing of Vietnamese positions in his country, provided Cambodians were not killed.[35]

After having permitted the North Vietnamese in, Sihanouk made the mistake of leaving the country to visit Beijing and Moscow. While he was away, he was deposed by Lon Nol and others, who then attempted to expel the North Vietnamese from Cambodia. They failed and suffered serious reverses at the hands of the far more numerous and battle-hardened Vietnamese. It wasn't American bombing that destabilized the Sihanouk regime; it was Sihanouk's weakness in the face of Vietnamese aggression.

In light of Hanoi's strength and fixity of purpose, as well as the determination of the Soviet Union and China to arm and support wars of "national liberation," none of this may have mattered very much at the end of the day. North Vietnam was determined to dominate the region. Without outside support comparable to that which North Vietnamese were receiving, the noncommunists could not hold on. And since the U.S. was ultimately unwilling to pay the price to stop the Communists, the die was cast.

In 1972, in accord with the agreement signed with the Americans, Vietnam withdrew most of its troops, leaving only the still weak Khmer Rouge to fight on. The Khmer Rouge bitterly resisted the cease fire and the agreement with the U.S. They regarded both as betrayals. Besides, as they themselves later acknowledged, if they had accepted a cease fire in 1972, their movement would have collapsed.[36] This admission certainly casts an interesting light on subsequent developments. If they were truly as weak as they believed themselves to be, it is quite likely that the Cambodian government would have prevailed against them—if Congress had not withdrawn all funding for the Indochina war.

Those who have claimed that the Khmer Rouge were somehow driven crazy by American bombing have to explain why the same did not happen to the Pathet Lao or the Vietnamese, who were bombed more heavily and for a much longer time than the Cambodians and yet did not commit genocide. Further, the Khmer Rouge did not actually take power until twenty months after the last U.S. bomb fell. As Peter Rodman observed, if the bombing drove them to genocide, it must have been a severely delayed reaction.

William Shawcross was not soft on Communism—at least not later. In 1994 he wrote scathingly of a British journalist, Richard Gott, who had been revealed as a spy for the Soviet Union. "I am astonished by the levity and frivolity with which his covert work for the KGB is being shrugged off by himself, by his editor and colleagues, even by some Tory luminaries," he wrote. Shawcross went further, acknowledging that he himself had once contributed to a climate of opinion about Communism that was dangerously wrong:

> In an interview, Gott describes his horrifying defence of Pol Pot and other such articles as "jeux d'esprit, pieces written against the tide." Yet while the Khmer Rouge was in power, murdering or causing the deaths of over a million people between 1975 and 1978 [estimates range between 1 million and 3 million], Mr. Gott was swimming against no tide; he was in the mainstream of the Left, many of whom either ignored the horror stories which refugees brought out of Cambodia or denounced them as CIA-inspired attempts to vindicate "bloodbath" theory....

The victims of Pol Pot's tyranny, or the Vietnamese victims of the brutal Hanoi regime which triumphed over the Americans in 1975, had little reason to thank Gott's "jeux d'esprit." Indeed those of us who opposed the American war in Indochina should be extremely humble in the face of the appalling aftermath: a form of genocide in Cambodia and horrific tyranny in both Vietnam and Laos. Looking back on my own coverage for the the *Sunday Times*, of the South Vietnamese war effort of 1970–1975, I think I concentrated too easily on the corruption and incompetence of the South Vietnamese and their American allies, was too ignorant of the inhuman Hanoi regime, and far too willing to believe that a victory by the communists would provide a better future.

But after the communist victory came the refugees to Thailand and the floods of boat people desperately seeking to escape the Cambodian killing fields and the Vietnamese gulags. Their eloquent testimony should have put paid to all illusions.[37]

Shawcross was rare—a journalist who learned from history and was willing to acknowledge his own errors. But he was virtually alone. It was perhaps because his book had come at such a convenient moment for the antiwar activists. Just when they were at their weakest—that is when confronted with the gruesome evidence of communism's human toll—*Side-show* relieved whatever guilt might have been flickering in their hearts. They were all too ready to swallow the idea that, really, when you examined the matter, it wasn't Communism that caused all that horror, but America—and the liberals' great nemesis—Richard Nixon.

CHAPTER FOUR

# The Mother of All Communists: American Liberals and Soviet Russia

*We cannot afford to give ourselves moral airs when our most enterprising neighbor [the Soviet Union] humanely and judiciously liquidates a handful of exploiters and speculators to make the world safe for honest men.*

—GEORGE BERNARD SHAW

THOUGH AMERICAN LIBERALS HAVE staunchly proclaimed their anti-communism in the decade or so since the hammer and sickle were consigned to the "ash heap of history," things were very different when the red flag flew over the Kremlin.

Certainly President Jimmy Carter did not sound like a "Cold Warrior" when he admitted after Soviet tanks rolled into Afghanistan in 1980 that "the action of the Soviets made a more dramatic change in my opinion of what the Soviets' ultimate goals are than anything they've done in the previous time I've been in office."[1]

President Carter had assumed the presidency proclaiming that the United States was now "free of that inordinate fear of Communism which once led us to embrace any dictator who joined us in our fear...."[2] In 1978 Secretary of State Cyrus Vance had told *Time* magazine that President Carter and General Secretary Leonid Brezhnev shared "similar dreams and aspirations" about the future of the world.[3]

To be sure, Nixon administration officials, avidly pursuing the policy of "detente" with the Soviet Union, had sometimes spoken a bit too fulsomely of "friendship" between our nations and other diplomatic folderol. But when President Nixon clinked champagne glasses with Soviet leaders, and opened the door to China with "ping pong diplomacy," he was practicing (wisely or not is a matter for historians) *realpolitik*, not naiveté. Though the detente policy caused dismay on the right—William F. Buckley Jr., for example, called the policy "impacted diplomatic hypocrisy...[that] has not achieved freedom for Eastern Europe...[nor] brought peace with honor to South Vietnam"—it was intended and perceived as a tactical maneuver.[4] Nixon ostentatiously balanced one Communist behemoth against the other—"playing the China card"—but this was not abandoning anticommunism as a guiding principle of American foreign policy.

For post-Vietnam liberals it was different. Anticommunism itself was considered discredited. The "opening to China" was hailed as a breakthrough for peace and right reason, not so much as shrewd diplomacy. It was virtually the only thing Richard Nixon ever did that was greeted by widespread liberal applause. And when Nixon made his visit to Beijing, the gushing reporting about Communist China was outsized. James Reston, a columnist for the *New York Times*, wrote:

> China's most visible characteristics are the characteristics of youth...a kind of lean, muscular grace, relentless hard work, and an optimistic, and even amiable outlook on the future....The people seem not only young but enthusiastic about their changing lives.[5]

Perhaps so many Chinese seemed young to Reston because so many of the older generation had been wiped out by Mao. But liberals traveling in China did not pause over such concerns. The economist John Kenneth Galbraith was even more impressed, saying that:

Somewhere in the recesses of the Chinese polity there may be a privileged Party and official hierarchy. Certainly it is the least ostentatious ruling class in history. So far as a visitor can see or is told, there is—for worker, technician, engineer, scientist, plant manager, local official, even, one suspects, table tennis player—a truly astonishing approach to equality of income.... Clearly, there is very little difference between rich and poor.[6]

American journalists traveling with President Nixon stressed the remarkable cleanliness of Beijing's streets, the absence of beggars, the elimination of crime, and the environmental benefits of bicycle transportation compared with the internal combustion engine. It went almost unremarked by members of the American press during Nixon's visit to China that the regime in power was directly responsible—through starvation, forced labor, and execution—for the deaths of 65 million Chinese.[7]

Meanwhile, at home, liberal members of Congress began a series of investigations into the military and intelligence agencies of the United States in hopes of discrediting them. Frank Church, a Democratic senator from Idaho, had absorbed the fashionable, leftist "lessons of Vietnam." In 1984 he described his foreign policy views to the *Washington Post*'s David Broder:

Remember how many years we pursued stupid policies in Asia, based on ignorance and an irrelevant ideological view of the world? The stupidity of it all! All those years of trying to "contain" China, a pygmy nation beset by problems of its own....

Until we learn to live with revolution, we will continue to blunder, and it will work to the Soviets' advantage. It will put them on the winning side, while we put ourselves on the side of rotten, corrupt regimes that end up losing.[8]

In a series of highly publicized hearings, Senator Church excoriated U.S. intelligence agencies, the FBI, and the United States Army for abuses. And there were some. The CIA had tested truth serums and psychotropic drugs on Americans without their knowledge or consent. One man, an army scientist named Frank Olson, had committed suicide in 1953 while unwittingly under the influence of LSD. The CIA had contrived or contributed to a number of coups d'etat, successful secret assassinations (mostly in South Vietnam),[9] and, in what became the Keystone Kops lead out of the weeks of hearings, a series of escalatingly bizarre assassination attempts against Fidel Castro—all of which, obviously, were unsuccessful.

Most Americans took these revelations more or less in stride, backing reforms and more oversight, but shying away from the extreme views of leftists who hoped to hobble the CIA entirely. When a CIA station chief in Greece was murdered after an anti-CIA group published his name, the public soured on shackling the CIA too severely.[10]

But for liberals, and this included many of the Democrats on the Church committee, the whole anticommunist thrust of American foreign policy had come to seem corrupt. Ronald Steel, writing in *Foreign Policy* magazine, spoke for many when he announced that Americans were "tired of the violence that has been committed in the name of peace . . . of the numerous interventions conducted in the tired vocabulary of anticommunism, of the sacrifice of their own unmet needs to an insatiable war machine, and of the deliberate deceit practiced by their leaders."[11]

## ANTI-ANTICOMMUNISM

Liberals did not, like hard leftists, defend or support the Soviet Union. They would acknowledge that the Soviet Union was undemocratic and repressive. Instead of outright support for the Communists— which, to be fair, they did not feel—they substituted contempt for what they always tagged as "primitive," or "knee-jerk," or "reflexive" anticommunism. The neoconservatives would later label them the "anti-anticommunists." Toward "cold warriors," and "right-wing anticommunists" the anti-anticommunists deployed a scathing scorn. Reaching for psychological explanations, they wondered why these warmongers were *seeking* an enemy in the Soviet Union. Later, when the Soviet Union had collapsed, the same people accused conservatives of "needing" an enemy to replace the old Soviet Union and finding it in places such as Iraq and China. Again, this was not so much an argument as political psychoanalysis.

The problem, through Carter's and other liberals' eyes, was not the existence of a communist threat but our groundless paranoia. After all, if our fear of Communism was once "inordinate," surely it was because the Soviet Union and other Communist states were not particularly threatening, nor particularly evil. Vietnam and its associated disillusionment had so soured liberals on America that they found it a terrible strain to condemn a communist nation. George Kennan, famous as the father of the containment policy, switched sides in the 1970s and joined what *Commentary* magazine later dubbed the "anti–Cold War Brigade." Kennan began producing articles about the Soviet Union's understandable fear of "encirclement" by NATO. The U.S., Kennan advised, "should follow a policy of minding its own business."[12] This was not so much isolationism as self-loathing. When Americans interfere in the affairs of other nations, ran the critique, they became "ugly Americans," and the nations receiving our attention suffered thereby. Right-wing isolationists of the 1930s had wished to keep America out of foreign entanglements on the grounds that we were too good for the world. Post-Vietnam liberal isolationism saw the world as too good for us.

Had ten new nations fallen into Communist hands between 1974 and 1980? Yes. South Vietnam, Cambodia, Laos, South Yemen, Angola, Ethiopia, Mozambique, Grenada, Nicaragua, and Afghanistan. Were nations from the horn of Africa to the Caribbean under siege by communist insurgents funded and supported by Moscow? Perhaps. But the Carter administration was untroubled by this. Zbigniew Brzezinski blandly described it this way:

> ...all the developing countries in the arc from northeast Asia to southern Africa continue to search for viable forms of government capable of managing the process of modernization.[13]

But as Jeane Kirkpatrick noted at the time, the peoples of these nations had absolutely no say as to what sort of government they would prefer. It was fatuous to suggest that these unfortunate victims of others' guns were "searching" for a way to "manage the process of modernization." They'd been captured. Secretary of State Cyrus Vance clearly did not see it that way. He implied that revolution of a Marxist variety was the inevitable march of history. "The fact is," he warned, "that we can no more stop change than Canute could still the waters."[14]

But certainly not even the most anticommunist American administration opposed "change." What they did reject were armed coups by small groups of well-armed and -supplied Communist insurgents who would seize illegitimate power and in every single case in history, go on to make war on their own people. No nation on the planet had ever chosen, through truly free elections, to be governed by Communists. And if they did hold phony elections, it was said to be "one man, one vote, one time." What Carter, Brzezinski, Vance, and the rest meant by "change" was actually the success of Communist coups d'etat. This theme arose again and again during Cold War debates. Despite the clear evidence of history that Communists had never won a free election, liberals persisted in the argument that they represented the popular will and took communist regimes at their word when they claimed to be pursuing the "people's" interests.

In a famous essay titled "Apes on a Treadmill" Jimmy Carter's director of the Arms Control and Disarmament Agency, Paul Warnke, summed up the post-Vietnam foreign policy of American Democrats and liberals:

> Neither we nor any other outsiders are wise enough to decide for another people the course to which their aspirations should lead them. The continuing penumbra of the illusion that somehow we know best can only blur a sound perception of our true foreign policy interests....
>
> The injection of American firepower into a local conflict is rarely compatible with our foreign policy interests. At a minimum, it will exponentially increase the devastation. A matching imprudence on the part of the other military superpower could engulf the world. Vaunting rhetoric about our peacekeeping role, our worldwide commitments, the morale of our allies, control of the seas, and our indispensable leadership of the free world now awakens as much derision as respect.[15]

Among academicians, two theories became popular at this time. One suggested that the systems governing the U.S. and USSR were "converging." They laid great emphasis on the bureaucratic nature of government in general, and of course, on a superficial level, this comparison was true, as far as it went. But it was utterly vapid. Both nations also had railroads; this did not imply that they were in any relevant respects "converging." When he ran for president in 1988, Massachusetts governor Michael Dukakis often demonstrated the influence of this school of thought. He certainly seemed to assume benign Soviet intentions. "There is nothing inevitable about conflict between the United States and the Soviet Union," he told the Peace Commission of the Episcopal Diocese of Washington. "...We should have begun long ago to work together on international problems—on health care; on cooperation in space; on world hunger, on the environment...."[16]

The other idea that gained currency proposed that the so-called North-South divide in the world, i.e., that divide between the haves and the have-nots, had become far more significant than the East-West divide. But as the close of the Cold War revealed, the USSR was itself an economic basket case, albeit one with an enormous military.

In the same speech in which he freed Americans from fear of communism, President Carter sounded his call for a foreign policy based upon human rights. What he did not say, but what the policy became in effect, was a program for lambasting America's noncommunist and anticommunist allies while permitting the monstrous human rights nightmare of the Communist world to go almost unmentioned.[17]

The invasion of Afghanistan changed things for Carter. The scales, he acknowledged, fell from his eyes. Yet those scales had proven remarkably durable.

Just during the time President Carter had been in office, the Soviets had tried and imprisoned Natan Scharansky (then called Anatoly Shcharansky) and Alexander Ginzburg merely for attempting to hold the Soviet Union to its promises in the Helsinki Agreement. Only a couple of years before Carter moved into the White House, the USSR had exiled Alexsandr Solzhenitsyn. The USSR had harassed, persecuted, and eventually sentenced to internal exile its leading physicist, Andrei Sakharov, father of the Soviet hydrogen bomb, as well as dissidents Yuri Orlov, Vladimir Bukovsky, and many, many more. These prominent human rights activists and refuseniks joined the less celebrated thousands of political prisoners occupying a concentration camp system dubbed the "Gulag Archipelago" by Aleksandr Solzhenitsyn. The Gulag, though reduced in scope from its heyday in the Stalin era—when it took the lives of some 20 million people—remained the world's largest penal system, and the world's most comprehensive effort to crush the human spirit.

Jimmy Carter may not have been paying close attention, but during the 1970s, the Soviets threatened to enter the Yom Kippur War if Israel destroyed Egypt's attacking forces, fomented revolution in a dozen countries, supported a network of provocateurs and spies within the United States, and circulated disinformation about America to the international press. They arrested Jews and Christians for the crimes of teaching religion or distributing Bibles, sentenced sane dissidents to mental institutions (presumably on the assumption that only a madman could fail to be blissful in the workers' paradise), and armed and supplied the deadly North Vietnamese, North Korean, Somali, and Ethiopian regimes. The Soviets encouraged Castro to send his soldiers to several African nations to serve as a Communist foreign legion, and supplied communist insurgencies in El Salvador, Nicaragua, and Grenada, among other places.

The Soviet Union during this period also engaged in one of the largest military buildups in the world's history, which included massive funding for chemical, biological, nuclear, and conventional weapons—in many cases in direct violation of signed treaties.[18] The USSR had supported and maintained a network of terrorists and spies around the globe. The infamous "Carlos," né Ilyich Ramirez-Sanchez, was a Soviet trained killer. The Soviets also gave funding and training to the Japanese Red Army, the Red Brigades, the Palestine Liberation Organization, the Baader-Meinhof gang, the French Action Directe, and the Popular Front for the Liberation of Palestine—all of whom made murder of innocent civilians a part of their political agitation.

All of this was only during the period of the 1970s. If President Carter had chanced to look into Soviet conduct stretching back to the 1930s, he would have found genocide on a scale equaled only by the Nazis and the Communist Chinese.

And yet President Carter misunderstood the true nature of that regime until its armies happened to cross a frontier on his watch. And really, until even after that. A glance at the rhetoric Carter and other liberals chose to use when speaking of the Soviet Union reveals, not the firm and consistent "containment" that so many would later claim, but instead a record of eager self-delusion and ready forgiveness of every Soviet aggression. From the debate over Vietnam, to the daggers-drawn contest over placing intermediate range missiles in Europe, to the protracted standoff on aiding anticommunist forces around the globe, liberals consistently misinterpreted the nature of the Soviet regime, misread its foreign and internal policies, and projected onto Soviet leaders their own fond wishes rather than realistic assessments. Sovietologist Jerry Hough, for example, could write in 1977 that "the Soviet leadership almost seems to have made the Soviet Union closer to the spirit of the pluralist model of American political science than the United States."[19] ABC reporter Walter Rodgers explained in 1986 that "Many Soviets don't want Western-style human rights, which they tend to equate with anarchy."[20] Dan Rather agreed: "Despite what many Americans think, most Soviets do not yearn for capitalism or Western-style democracy."[21] Mike Wallace of 60 Minutes was certain that the Russians really quite liked the boot on their necks: "Many Soviets, viewing the current chaos and nationalist unrest under Gorbachev look back almost longingly to the era of brutal order under Stalin."[22]

It seemed to pain liberals that attention to the Soviet Union's flaws, deceptions, and crimes inevitably made the United States look good by comparison. If there had been a way to condemn the Soviet Union without implicitly praising the U.S., liberals would have been more willing to do it. As it was, liberals could not entirely avoid acknowledging Soviet wrongdoing, but each and every time such a reckoning was pried from their unwilling lips—after the shootdown of KAL Flight 007, during the Shcharansky and Ginzburg trials, after Major Arthur Nicholson was shot in West Germany, and at other times—they rushed to add that the U.S. had certainly not been perfect in its international conduct; that we bore our share of the blame for tensions, conflict, and hatred; and that we must not presume to lecture others on morality until there was not a hungry child, poor adult, or illiterate migrant worker anywhere in the fifty states.

In time, this tendency to provide tit for tat examples of American wrongdoing in response to each and every episode of Soviet criminality hardened into a mindset of moral equivalence that was deeply demoralizing to the side that was, in all relevant respects, right.

## FIRST IMPRESSIONS

In the early days of the Soviet Union's existence, Americans and Europeans of liberal inclination swooned. *New York Times* reporter Walter Duranty, who was stationed in Moscow during Lenin's, and part of Stalin's, reign, offered this assessment of Lenin in 1921: "Lenin has a cool, far-sighted, reasoned sense of realities.... [He is willing] to put aside what experience has shown to be impractible theories and devote himself to rebuilding Russia on a new and solid foundation."[23] Duranty wrote of Josef Stalin, "Stalin is giving the Russian people— the Russian masses, not Westernized landlords, industrialists, bankers, and intellectuals, but Russia's 150,000,000 peasants and workers— what they really want, namely joint effort, communal effort."[24] In the 1930s, Duranty would play a critical role in debunking the "propaganda" that the USSR was experiencing a famine.

But, of course, the famine was quite real—not to mention intentional—and took the lives of at least ten million people. It was later called the "terror-famine" because it was the first time in modern history that any nation intentionally starved its own people. But readers of the *New York Times* heard nothing of it from their renowned Moscow correspondent. We know, through letters and other contemporaneous documents, that Duranty was well aware of the misery being imposed on Russia's peasants by the forced collectivization of agriculture. But his dispatches featured cheerleading accounts of the marvels of Soviet agriculture, stories that earned him the Pulitzer Prize "for dispassionate, interpretive reporting from Russia."[25] Only much later did his reputation suffer when it became impossible to deny that his reporting had been a skein of falsehoods.

Malcolm Muggeridge, the foreign correspondent for the *Manchester Guardian*, was the only Soviet sympathizer in the USSR who reported on the famine at the time. No prizes were issued to him. Instead he was often branded a liar. He wrote, "I had to wait for Khrushchev...for official confirmation. Indeed, according to him, my account was considerably understated. If the matter is a subject of controversy hereafter, a powerful voice on the other side will be Duranty's, highlighted in the *New York Times*, insisting on those granaries overflowing with grain, those apple-cheeked dairymaids and plump contented cows, not to mention [George Bernard] Shaw and other distinguished visitors who testified that there was not, and could not be, a food shortage in the USSR."[26] At the end of the twentieth century, it was learned that Duranty had been not a dupe but a blackmail victim. The Russians had apparently discovered his sexual proclivities and used this to control him.[27] This makes Duranty neither more nor less guilty historically, though it does suggest that he was not as stupid as those who swallowed Soviet lies credulously. Muggeridge, whose openness to the clear evidence before his eyes diminished his reputation in "progressive" circles, later rejected Communism completely.

Among those who were not blackmailed by the Soviet secret police, but wrote as if they were, was famed American novelist Theodore Dreiser, who penned worshipful reports of the egalitarian society he thought he saw. Soviet leaders, he marveled, earned no more than "225 rubles per month" while "among the Communist workers I could not find any who were earning less than 50 rubles and thousands upon thousands who were receiving 150, 175, 200, 225 or more!"[28]

Literary critic Edmund Wilson, too, was beguiled by what he imagined was true egalitarianism and joint effort. "Here the people in the park do really own it and they are careful of what is theirs. A new kind of public conscience has come to lodge in these crowds."[29] Wilson later called the USSR the "moral light at the top of the world."[30] Corliss and Margaret Lamont (Corliss was head of the American-Soviet Friendship Committee), popular writer H. G. Wells, Julian Huxley, Beatrice and Sidney Webb, American vice president Henry Wallace, Owen Lattimore, and many, many more were swept off their feet by the early Soviet Union. The Webbs may have been the most fulsome. In a book entitled *The Truth About Soviet Russia* published in 1942, they declared staunchly that Stalin was not a dictator but declined to make the same defense of Franklin Roosevelt. As for the Communist Party, it was, they wrote, "the most inclusive and equalized democracy in the world."[31]

Just before the announcement of the Hitler-Stalin pact that unleashed the Second World War (and sent honest leftists into a tailspin of anguish), a group of American intellectuals denounced their fellow liberals for being anticommunists. Among those who put their names to a full-throated defense of Stalinism published in the *Nation* magazine were Dashiell Hammett, mystery writer and longtime paramour of playwright Lillian Hellman; Harry Ward, chairman of the American Civil Liberties Union; I. F. Stone, the "gadfly" journalist; columnist Max Lerner, and four hundred others.

I. F. Stone was a lifelong leftist who used his biweekly newsletter to boost Castro, defend the Soviet Union, and condemn the United States. His biographer wrote that Stone saw "communism [as] a progressive force, lined up on the correct side of historical events."[32] And Stone himself admitted that he was "half a Jeffersonian, half a Marxist." The Marxist usually got the better of him.

When Stone died in 1989, he was hailed as the "conscience of investigative journalism" by the *Los Angeles Times*.[33] The *New York Times* obituary called him a "pugnacious advocate of civil liberties, peace, and truth."[34] TV personality Larry King called Stone "a truly genuine hero."[35] Both Anthony Lewis and Tom Wicker eulogized him in their *New York Times* columns, and Peter Jennings offered an on-air encomium, calling Stone "a journalist's journalist," and recommending his work: "For many people, it's a rich experience to read or re-read Stone's views on America's place in the world, on freedom, on the way government works, and sometimes corrupts."[36]

Sidney Hook, a tireless battler for liberty and democracy and scourge of American Stalinists, died within days of Stone. But his death went unlamented by American liberals. If liberalism were truly about respect for liberty, individual rights, and democracy, then it was Hook—not Stone—who exemplified those values. Stone's motto was "pas d'enemies á gauche" (no enemies to the left), and much the same can be said of his many admirers.

Stalin's defenders, responding to critics who insisted that the Soviet Union, like Nazi Germany, was a totalitarian regime, asserted that "the Soviet Union has established nationwide socialist planning, resulting in increasingly higher living standards and the abolition of unemployment and depression." The USSR, they believed, had complete intellectual freedom. "The best literature from Homer to Thomas Mann, the best thought from Aristotle to Lenin [Lenin!], is available to the masses of the Soviet people." And finally, "the Soviet Union continues to be a bulwark against war and aggression." Ten days later, the "bulwark" signed a pact with Nazi Germany and invaded Poland. Open admiration for the USSR became, for a time, unfashionable.

But it was never altogether lost. It might be thought that in those early days, particularly since accurate reporting from inside the USSR was hard to come by, liberals and leftists who made excuses for the USSR were simply ill informed. But in fact the 1930s and early 1940s were characterized by vigorous and spirited debates within left-wing circles about the nature of Communism. Figures of the anticommunist left like Sidney Hook, John Dewey, Lionel Trilling, and the members of the Committee for Cultural Freedom, which included dozens of writers, artists, teachers, and other intellectuals, provided ample published evidence about the arbitrary arrests, disappearances, and persecutions being carried out in Soviet Russia, as well as copious proof that organs like the New York Times were not telling the truth about the famine and other matters.

The evidence, in other words, was there for those with eyes to see. But in the 1930s and 1940s, no less than in the 1970s and 1980s, many on the liberal Left suspended their skepticism where Communist regimes were concerned. Even among those who were not actual members of the Communist Party, the Communist idea held hypnotic appeal. Neither the Purges, the Hitler-Stalin Pact, nor Stalin's brutal subjugation of Eastern Europe could break its spell. They clung to the Communist ideal with religious fervor and tenacity. As John Meynard Keynes put it in 1934, "Communism is not a reaction against the failure of the nineteenth century to organize optimal economic utput. It is a reaction against its comparative success. It is a protest against the emptiness of economic welfare, an appeal to the ascetic in us all.... It is the curate in [H. G.] Wells, far from extinguished by the scientist, which draws him to take a peep at Moscow.... The idealistic youth play with Communism because it is the only spiritual appeal which feels to them contemporary."[37]

There was a stark contrast between the imagined glories of the New Soviet Man and the realities on the ground for ordinary people under this comprehensive tyranny. The Soviet Union was not so much a state as a vast criminal conspiracy. Alexsandr Solzhenitsyn, Vladimir Bukovsky, Natan Sharansky, and others are the great chroniclers of the grotesque inhumanity of the Gulag and Communist rule. Robert Conquest, Walter Laquer, Richard Pipes, Merle Fainsod, Leonard Schapiro, and the editors of the *Black Book of Communism* have provided chapter and verse on the mass murders, deportations, political persecutions, abuse of psychiatry, and other depredations committed by the Communists. Side by side with political repression, the Soviets practiced a stubborn economics based on a false ideology that left the people perpetually undernourished, ill-clothed, and poorly housed. And now, with the opening of some sections of the Soviet archives, even more documentation about the lives of ordinary Russians, Ukrainians, and others has come to light. Typical is this letter sent to *Pravda* in 1932. It was never published:

> Comrade Editor,
>     Please give me an answer. Do the local authorities have the right forcibly to take away the only cow of industrial

and office workers? What is more, they demand a receipt showing that the cow was handed over voluntarily and they threaten you by saying that if you don't do this, they will put you in prison for failure to fulfill the meat procurement. How can you live when the cooperative distributes only black bread, and at the market goods have the prices of 1919 and 1920? Lice have eaten us to death, and soap is given only to railroad workers. From hunger and filth we have a massive outbreak of spotted fever.[38]

Writing years later (1973), Princeton professor and television pundit Stephen Cohen (a favorite guest on the *MacNeil/Lehrer NewsHour* in the 1980s) would write "Lenin's New Economic Policy, like Weimar culture, was a major chapter in the cultural history of the twentieth century, one that created brilliantly, died tragically, but left an enduring influence. . . . It was this toleration of social diversity, as well as the official emphasis on social harmony and the rule of law, as opposed to official lawlessness, that thirty years later would commend the New Economic Policy (NEP) to Communist Party reformers as a model of a liberal communist order, an alternative to Stalinism."[39]

Among the policies pursued by the Soviet Union under Lenin's New Economic Policy was equating absence from work with "sabotage." Under the NEP coal production increased fivefold within just one year. A triumph, Cohen might say. Of course, absence from work for any of the 120,000 coal workers was punishable by deportation to a camp or even death. In 1921 eighteen miners were executed for "persistent parasitism." Work hours were increased, including on Sundays, and the workers were so ill-clothed that men would often have to give their boots to members of the next shift. When they complained of hunger, commissars threatened to withdraw their ration cards altogether.[40]

Cohen imagines that Lenin's policies included a "toleration of social diversity." It's difficult to imagine what he meant by that. In May of 1922, Lenin sent a letter to Feliks Dzerzhinsky, head of the

Cheka (precursor of the KGB), describing the kind of "toleration" he expected in the new Soviet Union. The Cheka, he wrote, must impose "banishment abroad of all writers and teachers who have assisted the counterrevolution" [i.e. criticized the Bolskevik regime]. Further, "This operation must be planned with great care. A special commission must be set up. All members of the Politburo must spend two to three hours each week carefully examining books and newspapers.... Information must be gathered systematically on the political past, the work, and the literary activity of teachers and writers." Dzerzhinksy was diligent and began arresting and deporting hundreds of intellectuals and writers within a few months. But Lenin was impatient. He took up the matter with Stalin too, saying, "This issue came up in my absence. Has the decision been made yet to root out all the popular socialists? [Lenin provides some names.]...The Mantsev-Messing commission must draw up lists, and hundreds of these people should be expelled immediately.... The city needs a radical cleansing as soon as possible, right after the trial of all the Socialist Revolutionaries. Do something about all those authors and writers in Petrograd (you can find their addresses in *New Russian Thought* number 4, 1922, p. 37) and all the editors of small publishing houses too (their names and addresses are on page 29). This is of supreme importance."[41]

Even during the infamous Great Purge of the 1930s when Stalin rounded up and executed not hundreds, not thousands, but millions of Russians—including nearly all of the original Bolsheviks who had participated in the revolution—some leftists in the United States persisted in defending the Soviet Union. Upton Sinclair, author of *The Jungle*, explained why he believed the full confessions of "sabotage," "conspiracy," "treason," and more, signed by so many condemned Bolsheviks: "These men had withstood the worst that the Czar's police could do...and my belief is that the Bolsheviks would have let the GPU agents tear them to pieces shred by shred before they would have confessed to actions which they had not committed."[42] The United States ambassador to the USSR, Joseph Davies, was just as convinced that the trials were authentic. "To assume that this proceeding [Pyatakov-Radek trial] was invented and staged...would be to presuppose the creative genius of Shakespeare and the genius of Belasco in stage production."[43] Actually, it was more complicated than that, as Pyatokov was willing to do anything for the sake of the Communist Party. A decade before he had offered to shoot his own wife for her Trotskyite sympathies. And he was equally willing to confess to any crime if it would advance the cause.[44]

The devotion of American Communists was never tested in the same way, though any number might have proven themselves stalwart if asked. The folk singer Pete Seeger, whose popular image suggested irreverence and independence, was in fact a lockstep Stalinist. He crooned for peace from 1939 to 1941, when the Soviets and Nazis were allies. But when the Nazis invaded Russia, Seeger began to let his voice ring out for war—and for the "second front" Stalin was demanding. He sang odes to Stalin, and later to Ho Chi Minh.[45] Paul Robeson, the famed black singer and actor, was an eloquent spokesman for Negro rights, but he was also a dedicated Communist—which means that if his wishes had prevailed, no Americans, black or white, would have enjoyed civil rights. In 1948 Robeson traveled to the USSR after having said publicly that if there were ever a war between the U.S. and the Soviet Union, American "Negroes" would not fight for the United States. (Sugar Ray Robinson, the black professional boxing champion, said he would punch Robeson in the nose if he chanced to meet him.)

At the time of Robeson's journey, rumors had begun to circulate that Stalin was preparing a new pogrom against the Jews. Robeson announced publicly that while in Russia he would meet with Itzhak Feffer, a Jewish Communist poet who had visited the United States during World War II as a member of Stalin's Jewish Joint Anti-Fascist Committee. Feffer and Robeson had become friends. At the start of Stalin's anti-Jewish campaign, Feffer had disappeared. As David Horowitz recounts in his book *Radical Son*, "the question: 'What happened to Itzhak Feffer?' entered the currency of political debate."

When Robeson asked to see Feffer in Moscow, he was told he would have to wait, that Feffer was "vacationing in the Crimea."

In fact, Feffer had been in prison for three years and was so emaciated that he was near death. While Robeson waited, Feffer was released and given medical treatment. He was fattened up for several weeks, and only then permitted to meet with Robeson. Here is Horowitz's account:

> The two men met in a room that was under secret surveillance. Feffer knew he could not speak freely. When Robeson asked him how he was, he drew his finger nervously across his throat and motioned with his eyes and lips to his American comrade. "They're going to kill us," he said.
>
> "When you return to America, you must speak out and save us."
>
> After meeting with the poet, Robeson returned home. When he was asked about Feffer and the other Jews, he assured questioners that reports of their imprisonment were malicious slanders spread by individuals who only wanted to exacerbate Cold War tensions. Shortly afterward, Feffer, along with so many others, vanished into Stalin's Gulag.

It was not that Robeson had not understood Feffer's message. He had understood it all too well. Because it was Robeson, near the end of his own life and guilty with remorse, who told the story long after Feffer was dead.[46]

Another outright Stalinist, but one who never suffered pangs of conscience, and who was never adequately shamed during his long and highly prosperous life, was Armand Hammer. Hammer worked for the interests of the Soviet Union from Lenin's time to Gorbachev's. The FBI's file on him must have filled a whole drawer. And though among his other contributions to the Soviet cause was to serve as a paymaster for spies in the United States, he was never arrested, never tried, and in fact enjoyed the life of a wildly successful business tycoon.

Hammer is proof that even in the heady days of the Soviet Union's infancy, it was not only idealists and dreamers who were drawn to it. Hammer's father, a New York physician, was a Communist. At the age of twenty-four, at his father's bidding, Armand traveled to the USSR to meet Lenin. Lenin lavished contracts and other remuneration upon Hammer, recognizing the potential of an American businessman who was actually working for Moscow. One of the first deals Lenin gave Hammer was a concession for an asbestos mine in the Ural Mountains. In *Dossier: The Secret History of Armand Hammer* Edward Jay Epstein describes Hammer's impressions this way:

> Hammer was enormously impressed with the cold-blooded efficiency of the Cheka [the precursor of the KGB] and of its chief, Feliks Dzerzhinski. He later wrote approvingly about Dzerzhinski's personal intervention in railroad delays that same year. In considerable detail, he described an incident in which Dzerzhinski was on an inspection tour of the railroad in the Urals and cabled the local administrative center at Omsk to dispatch a train to pick him up. When the train did not arrive promptly, he went to the center with a detachment of Cheka troops and ordered the chief administrator and his assistant to step forward. They did, and were both taken to a courtyard and shot as a "lesson." ... Hammer pointed out that after the two lax officials were summarily executed, the trains ran more efficiently.[47]

## POST-VIETNAM ADJUSTMENTS

The Carter administration was the only presidency to test the anti–Cold War views of liberal Democrats. And it is quite possible that Jimmy Carter was able to perform this experiment only because of Watergate. In 1972, the Democratic candidate for president said, "I don't like Communism, but I don't think we have any great obligation to save the world from it. That's a choice other countries have to make."[48] Offered the chance to vote for a candidate whose slogan "Come Home America" was redolent of the Left's exhaustion with the Cold War, the voters decisively rejected George McGovern. Nixon's was the largest landslide in American history to that date.[49] And it is at least arguable that the American people were never willing to trust their nation to a candidate who failed to take a firm stand on national defense. In his 1977 autobiography, McGovern included pictures of himself in his World War II B-29, but also in a jeep with Fidel Castro, and with Vietnam's premier Pham Van Dong. His description of a 1976 visit to the Republic of Vietnam is a classic of liberal guilt:

> This visit to Vietnam moved me as has no other experience abroad. Walking through the streets of Hanoi where American bombs had fallen, seeing the devastation of the countryside from an airplane window, listening to Vietnamese expressions, despite all of this, of friendship for the American people redoubled my determination to work for reason and fairness in our postwar Vietnam policy."[50]

Of the nine presidents who served between 1946 and the collapse of the Soviet Union, four were Democrats, but only Carter was a post-Vietnam, anti-anticommunist liberal. (Though during the campaign of 1976, he had obscured this as much as possible, trading on his navy experience and his training as a "nuclear engineer" to give the impression of muscularity on defense.) He announced his intention to frame our relations with other nations within the rubric of human rights, but it was really an anti–Cold War foreign policy, and, as such, owed far more to McGovern than to Johnson or Kennedy.

# THE SOVIET UNION
# ACCORDING TO ANDREW YOUNG

President Carter's ambassador to the United Nations, Andrew Young, was the administration's most unbuttoned spokesman for the post-Vietnam worldview. And since his speeches hit all the high notes of the American liberal view of the Soviet Union, they are worth examining at length.

Speaking to a diplomatic audience in 1977, Young explained that the civil rights movement had changed America's foreign policy orientation "away from the tragedy of the past twenty years," when tax dollars had been spent "not to develop, not to feed the hungry, but essentially as part of an apparatus of repression in many places on the face of the earth."[51] This view, that the United States had been an agent of repression in the world, was certainly widespread on the Left. But it was extraordinary coming out of the mouth of a cabinet level officer of the U.S. administration. In point of fact, no nation had been as generous with aid and development assistance in the post–World War II period as the United States. Young further told his audience that not everyone in the U.S. had achieved post–Cold War enlightenment. He noted, with disapproval, that the U.S. was still spending "close to $100 billion a year on so-called military preparedness," and some of his fellow citizens persisted in speaking of a "so-called clear and present danger" which Young attributed to a "massive education campaign run by the government and private agencies."[52] Young invented a program that never existed to explain fear of a reality he denied.

While Young represented an administration that purportedly placed human rights at the very center of its approach to other nations, he was quite willing to understand and even rationalize the denial of human rights in the Soviet Union. Speaking at a New York church on "Human Rights Sunday," Young urged listeners to take a generous view of the Soviet Union's record on the matter:

> We must recognize that they are growing up in circumstances different from ours. They have, therefore, developed a completely different concept of human rights. For them, human rights are essentially not civil and political,

but economic.... One lives in a land where, in most of that land, the sun sets as early as three o'clock in the afternoon, and where the planting season is minimal. Under those circumstances the struggle for human rights inevitably becomes far more economic in its expression that it would in a country such as ours, where we almost take it for granted that anything can grow almost anywhere year 'round.[53]

Young was clearly not a farmer, nor conversant with the pioneer history of the American Midwest which featured generations of farmers struggling for survival in a country that may have enjoyed plenty of sunshine but also featured blinding blizzards, plagues of grasshoppers, floods, droughts, and prairie fires. Or he may have known that history and simply downplayed it out of deference to the Soviets' feelings. They, after all, had endured poor weather since, roughly, October 1917.

Whether Young was aware of it or not, he was parroting a staple of Soviet propaganda—indeed their most common rejoinder when accused of human rights abuses. The classic Soviet riposte was to contrast their emphasis on "economic rights" with our focus on political rights. Of course, the idea that a nation cannot "afford" political rights as it struggles to feed, clothe, and shelter its people is hardly self-evident. The reason the Soviet Union and every other Communist country that ever existed had trouble providing the people with the basics was allegiance to a false and utterly disastrous theory of economics. The truth, which liberals were slow to recognize, was that even if the Soviet Union and other communist states were judged purely on the grounds they preferred, namely the quality of life for ordinary people, they were vastly inferior to the free, prosperous West.

It may not have occurred to a single one of his listeners in that fashionable church in Manhattan that day, but there was something almost obscene about an officer of the United States government lecturing the

U.S. on sympathy toward the Soviet Union's farmers. For in doing so he was not actually seeking sympathy for ordinary Russians—a sympathy richly deserved—but toward the government that enslaved them. The peoples of the Soviet republics were not deprived because the growing season is short and the sun scarce in that part of the world (the country had been relatively well fed under the last tsar); they were deprived because their government had committed genocide against its farmers for more than a decade beginning in 1920 and stretching into 1933.

It began with severe taxes on the agricultural regions of the nation and then progressed to in-kind "requisitions," wholesale confiscations of property, and the "liquidation" of "kulaks" or so-called wealthy peasants. The Soviet regime's war against the peasantry did not start with Stalin. It began with Lenin, who in 1891 had praised the cleansing effects of rural famine, inasmuch as it would facilitate the creation of "a new industrial proletariat which would take over from the bourgeoisie. . . . Famine, he explained, in destroying the outdated peasant economy, would bring about the next stage more rapidly, and usher in socialism, the stage that necessarily followed capitalism."[54]

Once in power, Lenin was able to engineer famines himself.

World War I had devastated Russia's economy, and the civil war that followed further ravaged the nation. But it was Bolshevism that took the most lives. In 1920 Russia experienced a very bad harvest. Despite this, the Bolsheviks decreed that "all grain stocks, even the seed for future harvests, [be] seized."[55] The combination of taxes and confiscations provoked rebellion. Millions of peasants took up arms against their new government, though some could wield only pitchforks.

When famine ravaged the countryside in 1920 and 1921, international agencies as well as local Soviet officials pleaded with Moscow for assistance. But none was forthcoming. The peasants were regarded as enemies now, and would be starved into submission. The newly opened Soviet archives reveal this report to Moscow from an official of the People's Commissariat of Food from the Samara region:

> Today there are no more revolts. We see new phenomena instead: crowds of thousands of starving people gather around the Executive Committee or the Party headquarters of the soviet to wait, for days and days, for the miraculous appearance of the food they need. It is impossible to chase this crowd away, and every day more of them die. They are dropping like flies.... I think there must be at least 900,000 starving people in this province.[56]

In 1921 and 1922, five million people starved to death. Later, it would get worse.

One of the ironies of communism is that the stilted language it employed was so often the polar opposite of what it purported to describe. As George Orwell described so well in *1984*, those who made war called it peace, those who imposed ignorance called it strength, and those who imposed slavery called it freedom. Communists created the terms "crime against the people" and "enemy of the people." Nothing fitted those terms so well as Communists themselves. Stalin's forced collectivization of agriculture was a crime against the people of staggering proportions. The purported targets were "kulaks" or wealthy peasants, who were declared enemies of the people and hunted down like criminals. Though it is obvious that merely being a wealthy peasant ought not to have been a crime in any system, the term "wealthy" in this context was beyond absurd. Anyone above destitution could be denounced (often by a vengeful neighbor) as a kulak. The dekulakization brigades—teams of thugs with carte blanche to expropriate—had quotas to fill and did not perform careful inventories before striking. Between 80 and 90 percent of those persecuted as kulaks were actually middle-income farmers. The *Black Book of Communism* describes it this way:

> Peasants were arrested and deported for having sold grain on the market or for having had an employee to help with [the] harvest back in 1925 or 1926, for possessing two samovars, for having killed a pig in September 1929 with

the intention of consuming it themselves and thus keeping it from "socialist appropriation." Peasants were arrested on the pretext that they had 'taken part in commerce' when all they had done was sell something of their own making.... But most often people were classed as kulaks simply on the grounds that they had resisted collectivization.[57]

Eventually, the dragnet would include former policemen, priests, nuns, shopkeepers, relatives of officers in the White army, and numerous other "socially dangerous elements."[58]

The teams of commissars sent out to the countryside to confiscate the kulaks' property were in theory required to hand it over to the collectives. But it rarely happened that way. The Soviet collectivization set the pattern for what would later happen in China, Vietnam, Cambodia, Ethiopia, and many other nations. Communist officials or locals who bribed them got the privilege of pillaging their neighbors. A GPU (a forerunner of the KGB) report from Smolensk described how collectivization typically took place:

The brigades took from the wealthy peasants their winter clothes, their warm underclothes, and above all their shoes. They left the kulaks standing in their underwear and took everything, even old rubber socks, women's clothes, tea worth no more than 50 kopeks, water pitchers, and pokers.... The brigades confiscated everything, even the pillows from under the heads of babies, and stew from the family pot, which they smeared on the icons they had smashed.[59]

Between 1930 and 1931, nearly two million "kulaks" were deported. This was only a portion of those uprooted, executed, starved, shot, and sent to concentration camps to die slowly of disease, bitter cold, work, and starvation. As often as not, deportees did not make it to their destination. The *Black Book* offers an account by a Soviet official of the phenomenon of "abandonment in deportation." Writing to Stalin at a slightly later period and of a different category of victim—"outdated elements" from Moscow and Leningrad—the party official explained what happened to one group of six thousand people.

> On 29 and 30 April 1933 two convoys of "outdated elements" were sent to us by train from Moscow and Leningrad. On their arrival in Tomsk they were transferred to barges and unloaded, on 18 May and 26 May, onto the island of Nazino, which is situated at the juncture of the Ob and Nazina rivers. The first convoy contained 5,070 people, and the second 1,044; 6,114 in all. The transport conditions were appalling; the little food that was available was inedible, and the deportees were cramped into nearly airtight spaces.... The result was a daily mortality rate of 35–40 people. These living conditions, however, proved to be luxurious in comparison to what awaited the deportees on the island of Nazino (from which they were supposed to be sent on in groups to their final destination, the new sectors that are being colonized farther up the Nazina River). The island of Nazino is a totally uninhabited place, devoid of any settlements.... There were no tools, no grain, and no food. That is how their new life began. The day after the arrival of the first convoy, on 19 May, snow began to fall again, and the wind picked up. Starving, emaciated from months of insufficient food, without shelter and without tools,... they were trapped. They weren't even able to light fires to ward off the cold. More and more of them began to die.... On the first day, 295 people were buried. It was only on the fourth or fifth day after the convoy's arrival on the island that the authorities sent a bit of flour by boat, really no more than a few pounds per person. Once they had received their meager ration, people ran to the edge of the water and tried to mix some of the flour with water in their

hats, their trousers, or their jackets. Most of them just tried to eat it straight off, and some of them choked to death. These tiny amounts of flour were the only food that the deportees received during the entire period of their stay on the island. The more resourceful among them tried to make some rudimentary sort of pancakes, but they had nothing to mix or cook them in.... It was not long before the first cases of cannibalism occurred.[60]

Between 1932 and 1933, more than six million people from Russia, Ukraine, Kazakstan, and other agricultural regions, were executed outright or starved to death.

UN Ambassador Andrew Young wasn't finished. On the same day that he urged Americans to understand that climate was responsible for poor harvests in the USSR, he also praised their supposed elevation of economic "rights" and chastised America for its "reluctance to accept the concept of economic responsibility for all of our citizens."[61]

It was not until the very end of Soviet rule that this canard—that the Soviet Union, while certainly not free, at least provided a decent living for every man, woman, and child—was finally exposed. Before that there were jokes in the West[62] (and certainly in the East) about the long lines for basic goods, the poor quality of items manufactured in Communist regimes, and the terrible service, but the underlying mystique remained undisturbed. The Soviet Union was an egalitarian society—imperfect to be sure, but nonetheless a working model of Communism as envisioned by several centuries of philosophers. In the 1985 edition of his popular college text on economics, Paul Samuelson wrote this of the Soviet command economy: "What counts is results, and there can be no doubt that the Soviet planning system has been a powerful engine for economic growth."[63] Four years later, just before the Soviet Union went bankrupt, MIT economist Lester Thurow praised the "remarkable performance" of the Soviet economy and asserted that, "Today it is a country whose economic achievements bear comparison with those of the United States." Equally enthusiastic was Sovietologist Seweryn Bialer of Columbia University, who confidently asserted in 1981 that "The Soviet Union is not now, nor will it be during the next decade, in the throes of a true systemic crisis."[64] A decade after Bialer's prediction, the Soviet Union was dead.

Economist John Kenneth Galbraith wrote in 1984, "The Soviet system has made great economic progress in recent years.... One can see it in the appearance of solid well-being of the people in the streets." This, Galbraith explained, was because, unlike Western nations, the Soviet Union made "full use of its man-power."[65]

The truth was that the vast majority of the Soviet Union's citizens lived in deprivation and poverty, while a ruling class, or *nomenklatura*, had access to fine food, imported consumer goods, and other luxuries. The Soviet Union, westerners learned with surprise in 1991 and thereafter, was actually a Third World nation in terms of living standards. After the fall of the USSR, Moscow's health minister would reveal that half of the hospitals in the city had no sewerage, 80 percent lacked hot water, and some 17 percent did not have running water of any sort.[66] Conditions in the rest of the nation were infinitely worse. Following the Soviet Union's collapse, the true condition of public services, utilities, food, and sanitation began to get reported (though American journalists sometimes attributed the terrible state of things to the loss of Communism rather than to its disastrous seventy-year run). In 1992, the *Washington Post* reported that "gastrointestinal problems, dysentery, and salmonella [were] increasing" and "one-third of Russia's drinking water is unhygienic."[67]

Andrew Young did not find much of anything to criticize in the old Soviet Union. Asked about Cuban troops in Africa, Young was equable: "I don't believe Cuba is in Africa because it was ordered there by the Russians. I believe that Cuba is in Africa because it really has shared in a sense of colonial oppression and domination."[68] He further opined that Cuban troops brought "a certain stability and order to Angola."[69]

Young was America's ambassador to the United Nations when the Soviet Union tried Anatoly Shcharansky[70] and Alexander Ginzburg for treason. This was not, Young explained, evidence of the repressive nature of the Soviet government, but rather a "gesture of independence" (from whom he did not say) by Moscow. Besides, Young continued, "We also have hundreds, maybe thousands of people in our jails that I would call political prisoners." This was a bit much even for some liberals. Democratic senator Howard Metzenbaum of Ohio described Young's comments as "distasteful, intemperate and made without good judgment."[71]

But Young had many, many supporters, and when he was later obliged to resign his post for violating the U.S. ban on meetings with terrorist groups (he had met with a representative of the Palestine Liberation Organization), he became something of a martyr.

## THOSE SECRETLY LIBERAL SOVIET BOSSES

Throughout the Cold War, American liberals persisted in the hope that each new Soviet leader represented a triumph of reform and liberalization. The wish was father to the thought. W. E. B. DuBois wrote in 1937 that Stalin "asked for neither adulation nor vengeance. He was reasonable and conciliatory."[72]

When Yuri Andropov took power in 1982, a number of observers were optimistic. John F. Burns of the *New York Times* wrote:

> The diplomats predict that if Mr. Andropov has his way, which is far from certain, the first months of his tenure could see the beginning of a major crackdown on corruption, a cutback on some aspects of the bureaucratic red tape that entangles the Soviet system, and perhaps a move toward a slightly less rigid economic system.[73]

Princeton professor Stephen Cohen was not daunted by Andropov's career as head of the KGB:

Andropov seems to have been the most reform-minded senior member of Brezhnev's Politburo, an impression he chose to reinforce cautiously in his first policy speech as the new General Secretary. Nor does his 15-year stint as head of the KGB disqualify him as a potential reformer. Soviet police chiefs, who must understand the limits of control, have become advocates of liberalizing change before. Indeed, Andropov may be the only current leader who can assuage conservative fears of reform. And lest we forget that politicians sometimes rise above their former careers, Khrushchev once was called the "Butcher of the Ukraine" for his part in Stalin's terror.[74]

Cohen expects the reader to be comforted by this memory. But the comfort is cold. Khrushchev did acknowledge the scale of Stalin's crimes in the famous "secret speech" to the Soviet leadership, but that exhausts his reform credentials. He is also the Soviet leader who swore to "bury" the West, brought the U.S. and USSR to the brink of nuclear confrontation by placing nuclear ballistic missiles in Cuba, and sent tanks into Hungary to crush the reform movement there.

Dusko Doder, a reporter for the *Washington Post*, was no dupe when it came to the Soviet Union, and his dispatches were often quite valuable. But even in his case, the tendency to describe Soviet leaders as conventional, western-style democratic politicians instead of what they really were—gangsters—crept in:

From what little is known about him, the somewhat pro-fessorial-looking Andropov has been better prepared for supreme office than any of his predecessors. With the exception of Lenin, the founder of the Soviet state, he is the first intellectual in this post.

He likes theater and the arts and has written exten-sively on ideological matters. The 15 years he spent as head of the KGB security police have made him probably the most informed man in the country. In that post, he was, in a way, both foreign and interior minister in charge of a vast organization with foreign and domestic responsibilities.[75]

The head of the KGB must certainly be "the most informed man" in the country, though an expertise in other people's mail is hardly a qualification for high office. As for the idea that having an intellectual as a leader is cause for optimism, history doesn't suggest so. Among those who could fairly be described as intellectuals were Lenin, Trotsky, Pol Pot, and MaoTse-Tung. And to laud Andropov's skills as a manager on the grounds that he administered "a vast organization with foreign and domestic responsibilities" is like saying a drug dealer should be considered an expert on international trade.

Harrison Salisbury of the *New York Times* wrote that "The first thing to know about Mr. Andropov is that he speaks and reads English." This was later elevated in a different *New York Times* story to "fluent in English."[76] *Newsweek* reported that Andropov "spoke English and relaxed with American novels."[77] The *Economist* magazine also added a working knowledge of German to the general secretary's repertoire. But the *Washington Post* became truly carried away, reporting that Andropov was "fond of cynical political jokes with an anti-regime twist . . . collects abstract art, likes jazz and Gypsy music" and "has a record of stepping out of his high party official's cocoon to contact dissidents."[78]

All of these giddy descriptions of Andropov were debunked by Edward Jay Epstein in the *New Republic* magazine. The source for many of these accounts turns out to be one highly unreliable Soviet emigré who seems to have invented most of the details and who, in any case, left the Soviet Union in the 1960s. Then, quoting one another, journalists simply amplified the supposedly English-speaking Andropov into a polyglot, the jazz fancier into a devotee of Scotch whiskey. Harrison Salisbury's account of Andropov's English mastery, which was written in such a way as to give the reader the impression that Salisbury had witnessed it himself, was in fact, Epstein reports, based on a second-hand account by a "non-Soviet foreign visitor."

Years before *Time* magazine made Mikhail Gorbachev "Man of the Decade," they had made Ronald Reagan and Yuri Andropov joint "Men of the Year" in 1984. In classic moral equivalence style, the article began with two quotes—Reagan saying, "They are the focus of evil in the modern world"—and Andropov saying, "They violate elementary norms of decency." The lesson was clear enough: These leaders are two peas in a pod. Each makes extreme statements about the other. Neither is right.

The transformation of Andropov from KGB enforcer to Scotch-drinking, joke-telling, jazz-listening moderate, was for naught. He died just fifteen months after taking power. Still the hopefulness of the Western press corps flowered again for the elderly Constantin Chernenko. The *New York Times*'s John F. Burns wrote:

> Others caution against underestimating Mr. Chernenko, who impressed several Western leaders who met him after [Andropov's] funeral as a warmer, earthier man than Mr. Andropov, seemingly comfortable in his new role. Like Deng Xioping, who capped an even more remarkable comeback in China by launching radical reforms, Mr. Chernenko could still prove to be an old man in a hurry. If so, he will confound not only his critics but still more so those in the bureaucracy who look to him for a return to the less challenging times under Mr. Brezhnev.[79]

Sovietologist Jerry Hough had written about Chernenko in 1982, sounding the usual hopeful note:

> If Chernenko is purely a transitional leader, his policy preferences might not be very important. Nevertheless, if he does become the General Secretary, the coalition he has put together and the content of speeches and articles suggest

that he is a strong supporter of detente and of some kinds of reform. When Brezhnev presented Chernenko with awards on his 70th birthday, he praised his assistant for being "restless" in the good sense of the term, a man with a "creative, daring approach." In response, Chernenko acknowledged that he sometimes makes "nonstandard decisions." He has written often of the need for "further perfection of the political system," and frequently expresses what sound like anti-bureaucratic, pro-participatory views....[80]

When Brezhnev is your teacher and character witness, how reform minded can you really be? We'll never know. Chernenko, too, died before fulfilling his reform potential. If the sepulchral Constantin Chernenko was capable of stirring the hopes of gullible westerners, Gorbachev would prove their Bonnie Prince Charlie. As has been noted, they were prepared to swoon for any Soviet leader. And Gorbachev, unlike any of his predecessors, actually assumed power talking of reform, restructuring, and openness. His bywords—"perestroika" and "glasnost"—became better known in the United States than any Russian word since vodka. And the youngish Soviet leader became the object of quite rapturous coverage.

Professor Stephen Cohen, always on the lookout for a Soviet Communist reformer (and traduced by earlier hopefuls to the title) was certain that he had found his man in Gorbachev.

In campaigning for power and since becoming general secretary, he has pointedly associated himself with long-standing reformist ideas and constituencies in the Soviet establishment. Like earlier reformers, he has indicated that such a domestic program requires a relaxation of international tensions to increase non-military expenditures and to counter conservative protests that change is too risky.[81]

In other words, the U.S. must reduce defense spending in order to strengthen Gorbachev's hand at home. "The U.S. must decide," Cohen proclaimed, "whether it is a friend or foe of Soviet reform. A policy of cold war will almost certainly freeze any prospects of a Moscow spring."[82]

Serge Schmemamm, writing in the *New York Times Magazine,* zeroed in on his colleagues' predilections:

> There is something in the notion of young, educated, and smooth leader advocating change and lambasting the bureaucracy that the West finds irresistible. It is a feeling based on far more than wishful thinking—it draws on a deep-seated conviction that anybody pragmatic enough to see the obvious flaws of the communist system can only move his country closer to the Western world.[83]

CNN founder Ted Turner said, "Gorbachev has probably moved more quickly than any person in the history of the world. Moving faster than Jesus Christ did. America is always lagging six months behind."[84]

CNN's Moscow correspondent seemed to share his boss's enthusiasm. In a February 1986 letter to the *Wall Street Journal,* Stuart Loory wrote, "If suddenly a true, two-party or multiparty state were to be formed in the Soviet Union, the Communist Party would still win in a real free election. Except for certain pockets of resistance to the communist regime, the people have been truly converted."[85]

CBS's Bruce Morton, too, reported that the Soviet people were happy with Communism: "All of these services are part of an explicit bargain the Soviet workers have made with their government. They are less free than workers in the West, but more secure."[86] It might have been understandable (if not correct) to assert that Soviet workers had made an *implicit* bargain with their leaders. But an *explicit* one? In what forum was this bargain struck? Where was it spelled out? How was it monitored and enforced?

As for Gorbachev, NBC's Richard Threlkeld described him as "more polished than Nikita Khrushchev, but just as much of a populist reformer...a great communicator...and as cosmopolitan as Peter the Great."[87]

Dusko Doder of the *Washington Post* wrote glowingly of the new leader's "energy" and "vitality."

"His age alone was a factor. It immediately appeared to lift the nation's spirit.... Gorbachev's accession to power, other things being equal, holds out the prospect of a significant modernization of Soviet society in the coming decade."[88]

Mary McGrory, a columnist for the *Washington Post*, expected much more. Gorbachev, she gushed, had a "blueprint for saving the planet."[89] Dan Rather was swept away by the man's charm: "He has, as many great leaders have, impressive eyes...there's a kind of laserbeam stare, a forced quality you get from Gorbachev that does not come across as something peaceful within himself. It's the look of a human volcano, or he'd probably like to describe it as human nuclear energy plant."[90] Gail Sheehy wrote of Gorbachev's "luminous presence" and his "talent for reaching out to people of all social levels."[91]

A delegation of members of Congress visited the new Soviet leader and came away visibly impressed. Speaker of the House Thomas ("Tip") O'Neill was pleased with his "politician's informality" and also with "his solid grasp of the issues and of American politics." Comparing him to a "New York corporate lawyer," O'Neill found Gorbachev "easy and gracious." Massachusetts Republican Silvio Conte thought "he would be a good candidate for New York City... a sharp dresser...smooth guy."[92]

Bob Abernethy of NBC was quite sure that Gorbachev was equal to any challenge. Lauding him for rescuing the world from "suspicion and fear," he credited Gorbachev with holding "the first largely free elections here in seventy years." (Except that the Communists were not permitted to lose.) "Making this a truly modern country after years of tyranny is no easy task," Abernethy allowed, "but after all the other things he's done, here and throughout the world, one would have to conclude if anyone can do it, Gorbachev can."[93]

Sydney Blumenthal, later an advisor to President Clinton and scourge of conservatives, lauded Gorbachev as a great world figure who "created a starring role" for the clueless Ronald Reagan and thereby "rescued" the hapless fellow.[94]

It need hardly be said that it was inconceivable to imagine such encomiums being uttered by the same people about the American president, Ronald Reagan. In fact, Stephen Cohen was certain that Reagan was suffering from "a potentially fatal form of Sovietophobia."[95]

The content of Gorbachev's policies got little scrutiny—and could bear little. He did not rise to power repudiating or even criticizing the long history of repression, murder, coercion, and lies that were his political inheritance. Nor did he ever promise to make the Soviet Union more democratic, more liberal, or more "westernized." Instead, he pledged to make the country more "efficient." The problem, as Gorbachev analyzed it, was that the workers were drinking too much and working too little. His first "reform" was to limit the availability of vodka. His later reforms were equally unavailing because the problem was not a sound system that needed reform, but a totally corrupt and criminal organization that could never succeed. It could only be dismantled and replaced.

In order to lionize Gorbachev, liberals had to overlook quite a bit, such as his failure to reduce the Soviet Union's aggressive acquisition of weapons of mass destruction and his unflagging support for revolutionary movements around the world. Apologists claimed he was forced to do these things or risk the loyalty of the army. Perhaps. But what of small cruelties? In 1989, German chancellor Helmut Kohl, perhaps believing Gorbachev's clippings, begged him to release the hero Raoul Wallenberg, believed to have been yet alive in the Soviet Gulag. Gorbachev responded with silence.[96]

There is little doubt that Gorbachev was not the monster Stalin was; nor was he as rigid a Communist as many who preceded him. It is doubtful that any of his predecessors would have permitted Eastern Europe to liberalize. Gorbachev was the first Soviet leader who refrained from enforcing Communist rule by sending in the tanks. This may be why he was the last Soviet leader. The Soviet system was born in violence, committed grotesque atrocities while in power, and maintained itself through force and intimidation. When outright coercion was eschewed, the regime was finished. This was a thrilling development. But it did not earn Mikhail Gorbachev the status of secular saint. He was only a moderate bully in a system that required a vicious one.

In 1990, after an era in which Ronald Reagan had revived the American economy, restored the vitality of the western alliance, rebuilt America's world preeminence, and forced the Soviet Union to abandon its global ambitions, *Time* magazine made Mikhail Gorbachev Man of the Decade. Lance Morrow, in an essay that acknowledged the evil of the system Gorbachev hoped to reform, nonetheless favored him with the adjective "visionary... simultaneously the communist Pope and the Soviet Martin Luther, the apparatchik as Magellan and McLuhan. The Man of the Decade is a global navigator."[97]

Strobe Talbott had another agenda in that famous issue of *Time*. Not content to laud Gorbachev as the savior of the modern world, Talbott, in a gesture that proved just slightly ahead of its time, was eager to conclude that Gorbachev's new face of moderation meant that Communism was not and had never been threatening:

> For more than four decades, Western policy has been based on a grotesque exaggeration of what the USSR could do if it wanted, therefore what it might do, therefore what the West must be prepared to do in response....
>
> In order to believe the Soviet Union is capable of waging and winning a war against the West, one has to accept as gospel a hoary and dubious cliche about the USSR: the place is a hopeless mess where nothing works, with the

prominent and crucial exceptions of two institutions—the armed forces and the KGB.

Talbott had no patience for such cliches. The Soviet Union, he urged, was not expansionist and never really had been. Yes, Stalin had gobbled up Eastern Europe, but "he did so in the final battles of World War II, not as a prelude to World War III. The Red Army had filled the vacuum left by the collapsing Wehrmacht." In Warsaw, Budapest, and Prague? Talbott's charity regarding Soviet actions was not available for U.S. policymakers. "In its unrelenting hostility to Cuba, Nicaragua, and Vietnam," he wrote, "the Bush administration gives the impression of flying on an automatic pilot that was programmed back in the days when the Soviet Union was still in the business of exporting revolution." Talbott went on to counter claims by conservatives that Reagan's toughness and steady pressure had paid off in the changes taking place behind the Iron Curtain. No, Talbott insisted, "Gorbachev is responding primarily to internal pressures, not external ones."[98]

But those "internal pressures" were, at least in part, created by a deliberate policy of the Reagan administration. Never the cowboy he was caricatured as, Reagan applied a series of squeeze measures during the eight years of his leadership—policies that were later credited by the former Soviets themselves for providing the final shove that sent the Soviet Union tumbling onto the "ash heap of history." His administration restricted high technology and dual use exports to the Communist world, backed anticommunist insurgents as well as democratic resistance groups throughout the Soviet Empire, rebuilt America's conventional and nuclear forces at a clip the Soviets could not match without disastrous domestic consequences, and maintained a spiritual offensive by calling the system by its proper name—evil.

While Reagan's anti-Soviet rhetoric won him contempt from liberal opinion leaders in the West, it was greeted with relief and even joy in the Communist world. Bartak Kaminski, a Polish professor, said Reagan was "the first world leader of the post-detente era who was willing to express ideas about the Soviets that were shared by most Poles."[99]

Andrei Kozyrev, the first foreign minister of the Russian Republic, remarked just after the failed coup of 1991 that it had been a mistake to call it "'the Union of Soviet Socialist Republics.' It was, rather, an evil empire, as it was once put...."[100]

Oleg Kalugin, a former general in the KGB, and Yevgeny Novikov, a former senior staffer of the Soviet Communist Party Central Committee, confirmed that "American policy in the 1980s was a catalyst for the collapse of the Soviet Union."[101]

And former Soviet foreign minister Alexsandr Bessmertnykh told a post–Cold War gathering at Princeton University that "SDI made us realize we were in a very dangerous spot.... Gorbachev was convinced any attempt to match Reagan's Strategic Defense Initiative... would do irreparable harm to the Soviet economy."[102]

Though he intended it as the severest censure, leftist Richard J. Barnet got it pretty much right in describing the Reagan approach:

> In Soviet-American relations, the United States has signaled its intention to dash rather than build a relationship based on mutual interest. In Washington's view, our Soviet adversary is to be managed by steadily increasing the threat we pose to it. Indeed, the Administration appears to deny that we have any interest in common with the 'evil empire.'... Washington evidently prefers to rely instead on our own technological edge and superior economic power.[103]

Barnet is mistaken only in that the Reagan administration did not intend to "threaten" the USSR, merely to reassert American determination and values. The Reagan years reinvigorated the West militarily, economically, and spiritually. In the end, this challenge proved too much for a communist world grown weary and arthritic.

Very few on either end of the political spectrum had predicted the Soviet Union's abrupt collapse. The event was disorienting for everyone. But if liberals maintained anything, it was this belief: Nobody— least of all their old nemesis Ronald Reagan—had *won* the Cold War.

# CHAPTER FIVE

# Fear and Trembling

*A liberal is a man too broadminded to take his own side in a quarrel.*　　　　　　　　　　　　　—ROBERT FROST

WHILE IT IS CERTAIN THAT LIBERALS lost their taste for the fight against world Communism during the Vietnam War, they became even more hostile to the enterprise as their fear of nuclear war grew in intensity during the 1970s and 1980s. Their fear was aggravated by the election of Ronald Reagan in 1980. Reagan had cheerfully violated the sacred taboos of the liberal dispensation. During his campaign for the presidency, he described the Vietnam War as a "noble cause." At his first press conference from the White House, he asserted that the Communists had "openly and publicly declared that the only morality they recognize is what will further their cause, meaning they reserve unto themselves the right to commit any crime, to lie, to cheat."[1] And Reagan set about fulfilling his promise to increase vastly defense spending. All of this convinced the liberal establishment that Reagan was plunging the nation and the world toward nuclear catastrophe. And while even the Carter administration had occasionally been obliged to criticize the Soviet Union, none of the four presidents who preceded Ronald Reagan thought of themselves as fighting the Cold War. Though we now speak of the Cold War as having ended in December 1991, the term was used during the 1970s and 1980s only in the past tense, to refer to the Eisenhower-Kennedy years. The 1970s and 1980s were presumed to belong to a new period, that of "detente" or "peaceful coexistence."

It is difficult to overstate the degree to which Reagan was a departure from previous presidents. The arms control process was regarded by those presidents and by most of the opinion-shaping elite in the country as the most important aspect of the relationship between the two nations. Since the Johnson years, successive American administrations had accepted, to varying degrees, Robert McNamara's belief that nuclear superiority should not be the goal of U.S. policy; that we should strive instead for a "balance of terror" even if that meant permitting the Soviets advantages in certain areas. This doctrine would come to be known as MAD, or mutual assured destruction. But in this, as in so many other aspects of the U.S.-Soviet relationship, Reagan was determined to depart from the prepared script.

Reagan not only gave every indication of wanting to fight the Cold War; he also seemed to think the West would win it. In a speech to the British parliament in 1982, he displayed more prescience about the fate of the Soviet Union than was demonstrated by most Sovietologists, scholars, or CIA analysts:

> In an ironic sense Karl Marx was right. We are witnessing today a great revolutionary crisis, a crisis where the demands of the economic order are conflicting directly with those of the political order. But the crisis is happening not in the free, non-Marxist West, but in the home of Marxist-Leninism, the Soviet Union. It is the Soviet Union

that runs against the tide of history by denying human freedom and human dignity to its citizens. It is also in deep
economic difficulty.[2]

This kind of language embarrassed and infuriated many Americans and Europeans. Though Reagan's views are now retrospectively
claimed by everyone from Bill Clinton to Bill Bradley to Madeleine
Albright, they were considered dangerous and foolish at the time.
They also dismantled years of carefully built lies and distortions
aimed at persuading Americans that their country was as responsible
as the Soviet Union for tension, danger, and evil in the world. George
McGovern proclaimed:

> I think we ought to be very thankful that this man
> Andropov seems to be a reasonable guy and somewhat
> restrained. Because certainly the Reagan-Weinberger ap-
> proach is one of intense confrontation. It's almost as
> though they were spoiling for a military showdown.[3]

When Reagan's appointees testified on Capitol Hill, they were
routinely regarded with suspicion and even hostility by the Democrats. Secretary of State George Shultz, discussing U.S.-Soviet relations
before the Senate Foreign Relations Committee in 1983 found himself the target of highly tendentious questioning. Senator Charles
Percy of Illinois (a liberal Republican) sounded the familiar lament
that the Reagan administration had not yet held a summit meeting
with the Soviet leader. (Reagan later quipped about this, "They kept
dying on me.") Democrat Paul Tsongas of Massachusetts offered that
"everybody else thought there was some value to face-to-face negotiations." Why was the Reagan administration resisting? Alan
Cranston, the California Democrat who would run for president as
the nuclear freeze candidate the following year, demanded to know
what "the United States, for its part, has done to contribute to the tension that exists between the United States and the Soviet Union."
Shultz, to his everlasting credit, replied: "Nothing."[4]

The Democrats in Congress did not believe in fighting, far less winning the Cold War. Scarred by Vietnam (for all the wrong reasons) and tolerant of Third World flirtations with communism, they believed above all in one thing: arms control. It was the sheer ownership of weapons, they argued, that endangered the peace of the world. Wars, they believed, could not be avoided when nations engaged in "arms races." Particularly in the nuclear age, the argument went, miscalculation or accident could set off Armageddon. They further believed that nations armed themselves and prepared for war only because they labored under "misunderstandings." Disarmament, therefore, was the road to sanity and peace, while those who urged upgrading or improving America's military capacities in any way were Dr. Strangeloves, blind to danger and possibly mentally unbalanced. Disarmament was a sacrament to liberals. They were for it—mutually if possible—but unilaterally if necessary. Those who opposed disarmament were condemned as enemies of "peace."

Reagan responded to this fantasy by quoting Salvador de Madariaga, chairman of the League of Nations Disarmament Commission: "We do not distrust each other because we are armed; we are armed because we distrust each other."[5] It was not the weapons, he argued, but the nature of regimes that endangered peace. Both the U.S. and Canada, Reagan and other disarmament skeptics pointed out, were heavily armed but feared nothing from one another. Nor was it the case that nations always armed themselves only in response to the actions of their adversaries. Liberals invoked this image time and again, with varying metaphors—apes on a treadmill, scorpions in a bottle—and it became a truism that the arms race had a dynamic of its own with each side reacting to escalation by the other in a dangerous spiral that could end only in conflagration. This was patently untrue. The U.S. responded not just to the Soviet Union's massive arms buildup in the post–Cuban Missile Crisis years, but also to its internal repression, its support for revolutions and terrorists worldwide, its lies and fierce anti-American and anti-Semitic propaganda, and much more.

Anthony Lewis of the *New York Times* was often the perfect distillation of liberal opinion on these matters. His column, "Abroad at Home," was a bellwether of progressive opinion. Here is a sample of a 1983 Lewis column that deserves quoting at length:

> The Reagan policy for dealing with the Soviet Union—bristling words and an all-out arms race—is flourishing as never before.
>
> The Administration has succeeded in using the Korean airliner incident to intensify anti-Soviet rhetoric and win Congressional approval of an array of new weapons. But... at the moment of its fullest application, the policy has proved bankrupt. A number of Administration officials, realizing the dangers, are alarmed. Even Ronald Reagan may have an inkling.
>
> The event that put a chill on Washington was the statement by Yuri V. Andropov on relations with the United States.... The Soviet leader... implied that there was no point in trying to do business with such a government.
>
> ... It is one thing to give sermons about the evil of the Soviet system [which Lewis had vigorously condemned]. It is quite another to wake up and realize that the leaders of the other superpower no longer think it worth talking to you about matters engaging the fate of mankind. If meaningful communication breaks down, the consequences could be immediate and severe on the question of medium-range nuclear missiles in Europe.... Once the deployment of Pershings and cruise missiles begins, the Russians may take some threatening action in response.

Lewis went on to predict nuclear submarines moving closer to the East Coast of the United States and Soviet missiles in East Germany.

## KAL 007 AND THE NATURE OF THE REGIME

What Lewis called the "Korean airliner incident" refers to the decision by the Soviet military command to shoot down an unarmed, civilian airliner that had strayed into Soviet airspace in August of 1983. South Korean Airlines Flight 007 was cruising at 35,000 feet on its way from Anchorage, Alaska, to Seoul, South Korea. For two-and-a-half hours, Soviet military jets had shadowed the huge passenger liner. The fighters did not attempt to contact the passenger plane by radio, nor to force it down through visual contact. Instead, on orders from the ground, the Soviet fighters blew the airliner apart just minutes before it would have left Soviet airspace. KAL 007 had been carrying 269 passengers and crew including a member of the U.S. Congress. There were no survivors.[6]

At first, the Soviet Union denied any knowledge of what had become of Flight 007. But after five tense days of mounting American outrage, Foreign Minister Andrei Gromyko finally acknowledged that the Soviets had shot the plane down. "We state," he declared icily, that "Soviet territory, the borders of the Soviet Union, are sacred."[7]

It was a graphic illustration of the nature of the regime—and one that could not be easily relegated to the inside pages of the newspaper. President Reagan responded with outrage, calling the shootdown of an unarmed passenger airliner an "atrocity." The United States embarrassed the Soviets at the United Nations by playing the tapes of commands from the ground to the pilots to "destroy the target."

For most Americans, the shootdown of KAL 007 was illustrative of the enormous moral gulf that separated us from the Soviet Union. The Soviets' first story was complete denial. When this evaporated under the glare of hard evidence, they offered the excuse that they mistook the 747 for a spy plane with a similar configuration. This is difficult to credit, since the 747 has an unmistakable silhouette—even to a layperson. And even if the Soviets were sincerely confused, their decision to shoot down the intruder without so much as attempting to contact the plane or wave it off was barbarous.

But as chilling as the shootdown was for most Americans, many liberals worried more about its effect on the U.S. posture toward arms control than upon its true meaning. David Corn, a writer for *Nuclear Times* magazine, and later an editor at the *Nation*, penned a lament to Soviet leaders, regretting that:

> The incident feeds [sic] into the hands of cold warriors, who have been quick to appeal to the press with demands for a tough response—for ending normal diplomatic relations and canceling grain sales, for blocking Aeroflot landings here and even halting arms negotiations. Here, they insist, is evidence of the true nature of the Soviet regime.[8]

For some, like the editorial writers at the *New York Times*, the incident was only further proof that arms control negotiations were urgent:

> President Reagan's decision not to let the Soviet attack on a Korean airliner disrupt arms control talks was a courageous rebuff to some of his conservative allies. New evidence of Soviet paranoia only strengthens the case for curbing the arms race and maintaining a stable military balance. The question now is whether the aim is talk or agreement. For real progress, the President has to seek difficult compromises.[9]

In other words, a barbaric act by the Soviet Union strengthened the case for American capitulation to Soviet demands in arms control negotiations. It is hard to imagine what, in the eyes of the *New York Times*, could possibly have *damaged* the case for American concessions.

In Lewis's treatment, the "incident" is merely the occasion for Reagan to "intensify" his "anti-Soviet rhetoric." This is the sort of anti-American jujitsu for which Lewis was renowned. The trouble was not Soviet barbarism, but as Lewis describes it, the cynical exploitation of the episode by the anti-Soviet president. And now

President Reagan had made the Soviet leader so angry that he was threatening not to engage in arms control negotiations!

Lewis further predicted that if the Americans persisted with plans to deploy cruise and Pershing II intermediate range nuclear missiles in Europe, the Soviets would respond with missiles in East Germany. But what Lewis neglected to mention is that the Soviets had already placed their own intermediate range nuclear missiles, the SS-20s, in Europe. The NATO deployment of Pershing IIs was a response to the Soviet move.

In the late 1970s, in part due to the Carter administration's passivity, the Soviets had been emboldened to alter the traditional balance of power in the European theater. Until then, overwhelming Soviet superiority in conventional forces had been balanced by American superiority in nuclear weapons. The free nations of Europe were said to be protected by an American "nuclear umbrella." By placing intermediate range missiles in Europe that could reach all of the major capitals in as little as six minutes, the Soviets altered the calculus of deterrence, just as their earlier attempt to put missiles in Cuba would have shifted the balance of power. Leaders of Germany, Britain, and other NATO allies begged the United States for a response, in the form of Pershing IIs.

It was actually the Carter administration that had first made the decision to deploy the Pershings. But at that time (1979), the missiles were not ready. It fell to Reagan to implement the decision and he was forced to do so against a backdrop of escalating nuclear paranoia (egged on by Soviet disinformation and propaganda) in the West. It was a delicate moment for the NATO Alliance, as the Soviets did everything in their power to drive a wedge between the U.S. and its European allies. A failure to respond to the Soviet SS-20s might have spelled the end of NATO.

The governments of Western Europe were under terrific strain from their own pacifist wings. In April of 1981, fifteen thousand West Germans took to the streets of Bonn to protest the deployment of

NATO missiles. In November, a quarter of a million Germans marched, as did similarly sized throngs in London, Paris, Brussels, and Milan.[10] Protesters wore death masks and carried coffins. In Germany, they carried signs reading "Better neutral than dead," and, "We don't want to be defended to death!" Another read, "I don't know if the Russians will come. But I know the Amis [Americans] are here."[11] Willy Brandt urged acceptance of the Soviet offer to reduce the number of its SS-20s if the U.S. would forego deployment of the Pershing IIs. "Why haven't we," he asked the cheering crowd, "taken the Soviets at their word in Geneva?" Arlo Guthrie sang "Blowin' in the Wind." Similar, if smaller, demonstrations took place in more than 140 U.S. cities. In the U.S., mothers marched with babies in strollers carrying placards with slogans such as, "You can't hug your kids with nuclear arms."[12]

The Soviets were busily attempting to fan these flames through propaganda and so-called "active measures," a KGB term for political manipulation. The Soviets also employed crude threats. Just before West Germans went to the polls in 1983, visiting Soviet foreign minister Andrei Gromyko said, "In the nuclear age, the Federal Republic of Germany and the Soviet Union are, figuratively, in the same boat. If there are gamblers and con men who state that they are ready to plunge humanity into a nuclear catastrophe for the sake of their ambition, who gave them the right to put the people who want to live down into the abyss with them?"[13] (The threat was unavailing. The West Germans elected the Christian Democrats, who supported the Pershing II deployment—not the Social Democrats, who opposed it.)

In Western Europe, the Soviets tended to hand off the political portfolio to local Communist parties. A briefing paper prepared by the West German interior ministry described Soviet behind-the-scenes maneuvering:

> In the Federal Republic [West Germany] the Soviet Union has at its disposal for the advancement of its goals the German Communist Party (D.K.P.) and the linked organizations

ready to serve it. The D.K.P. acts either in its own name, or in those of its related or influenced organizations, to bring the interests and directives of the Soviet Union into the planning and content of the actions of the peace movement in the Federal Republic.[14]

The Swiss government expelled the Bern bureau of the Soviet news agency Novosti on grounds that it was "grossly interfering" in Swiss affairs, specifically in the antinuclear movement.[15] As Stanislav Levchenko, a KGB defector, told the *New York Times* in 1983, most of the antinuclear activists were completely honest and above board, but:

> ... They want a leader or two. They want somebody who stays late to write out the platform when they go home to bed. Those people stay busy. Sometimes it's just a slogan. But the degree of Soviet success so far has been great. The buildup of criticism on nuclear weapons by these groups has gone basically in only one direction—against NATO.[16]

This was true of all aspects of the "peace movement." The agitation to disarm or to compromise or to capitulate was utterly one-sided—all of the pressure was on the West and the free governments of Europe. No one, as Irving Kristol observed at the time, stopped to ask, "Why did the Soviets insert those SS-20s in the first place?"

> They didn't have to, if it was merely military security they were concerned about. The Russians already had a preponderance of power in Europe, a preponderance of nuclear as well as conventional power. So why did the Soviet leaders decide to do what they must have known would be exceedingly provocative—namely, install these new, very powerful, very accurate missiles? The only possible explanation is that the Russians are not satisfied with having a clear edge in the balance of power in Europe; they

want an overwhelming preponderance of power over Western Europe. They want Western Europe to be radically inferior militarily—radically vulnerable—to the Soviet Union. One can further assume that if they wish to achieve such overwhelming superiority they intend to do something with it, like intimidating the nations of Western Europe to pursue policies, economic and political, which are skewed toward the interests of the Soviet Union.[17]

In his memoir, former Soviet ambassador to the U.S. Anatoly Dobrynin acknowledged what few American liberals ever would admit: that the placement of the SS-20s—not the American response—set off "another sharp spiral in the strategic and conventional arms race."[18] The Soviet Union undertook this provocation, Dobrynin further admitted, in spite of the fact that "we were already well ahead of the United States in the number of armored vehicles, artillery pieces, and self-propelled launchers but behind in some strategic weaponry."[19]

There was no issue that aroused greater passion than arms control among American and European liberals and leftists. To be labeled an "opponent of arms control," as Assistant Secretary of Defense Richard Perle was on a routine basis, was to be branded as an opponent of simple decency and common sense. By such means did liberals routinely pressure their governments into arms negotiations. The democratically elected governments were thus forced to prove their peaceful intentions (to their own domestic constituencies), while the tyrannical Soviet Union, having no domestic constituency to placate, could simply push for its own advantage.

And the USSR succeeded. Nearly every arms accord favored the Soviet Union. SALT I, for example, permitted the Soviets to build more ICBMs than America was allowed. By 1982, the Soviets enjoyed a 1.63 to 1 advantage in strategic forces (ICBMs, long-range bombers, and SLBMs), and a 3.68 to 1 advantage in throwweight.[20] The Soviets certainly viewed SALT I as a victory. Georgi Arbatov of the Soviet Institute on the USA, bragged in 1972 that the U.S. had signed SALT I because "new...conditions" had required it. Among these new conditions were "the increased might of the Soviet Union, American setbacks in Vietnam, and such American domestic problems as unemployment, currency difficulties, inflation, and race."[21]

Though arms control agreements favored the USSR, the Soviets flouted them anyway. Reagan had not exaggerated: They violated nearly every arms agreement they had signed, and the violations of the ABM Treaty and SALT II were particularly glaring. But American liberals frequently found benign interpretations of Soviet violations— when they acknowledged them at all.

Members of the Center for Defense Information and its president, former admiral Gene La Rocque, were frequent guest experts on PBS's influential *MacNeil/Lehrer NewsHour* and other public affairs programs and were often quoted in major news stories. Because the group was led by an ex-admiral and staffed by other former navy men (most of whom had known La Rocque in the service), their soft, anti-defense-spending posture was particularly sought after on liberal-leaning shows. Here is what the center's newsletter had to say in 1987 about the Soviet Union's Krasnoyarsk phased-array radar, which the Reagan administration charged was a violation of the ABM treaty:

> The administration believes that it is an early warning radar because it closely resembles other Soviet radars of this type. If so, it would be in violation of the ABM treaty.... The signals emitted by the radar when it is turned on will help to resolve the issue.... In the meantime, the Soviets reportedly offered to halt construction on the radar if the U.S. would do the same on two early warning phased-array radars that it is installing in Greenland and the U.K. Whether the U.S. explored the offer is not known, but it seems clear that the adminstration's high voltage

publicity... is not calculated to encourage the Soviets to push their offer further.[22]

Though the language is deceptively mild, the CDI was both winking at a Soviet treaty violation and suggesting that the United States halt a legal operation in order to get the Soviets to halt an illegal one. If the United States *had* made such an offer, it would have been an obvious reward to the Soviets for cheating. Further, if the United States had been the one caught cheating, groups like CDI would have been, as it were, up in arms. Following the USSR's demise, former Soviet foreign minister Eduard Shevardnadze admitted that the phased array radar at Krasnoyarsk was an "open violation" of the ABM Treaty.[23] Years later, when the Soviet Union had dissolved, Admiral La Rocque was still nervous about U.S. military power. At a reception honoring the leftist Institute for Policy Studies in 1993, he said, "There's a strong strain of bellicosity and jingoism and chauvinism rampant in our country. We're very comfortable with war. You might even say we like war. I would go so far as to say we love war in this country."[24]

Reagan would eventually consider sweeping arms control proposals with Mikhail Gorbachev, but he entered the presidency dubious about such treaties, convinced that they had not historically advanced U.S. interests. Though he was reviled and ridiculed by many Democrats for it, Reagan believed that peace could best be achieved through American strength—military, political, and moral. The political and moral dimensions of the struggle became clear over time; the promise to reverse the slide in U.S. military preparedness had been a campaign promise.

## ABOUT THAT ARMS RACE

Despite the cliché of a thousand editorials, it was not correct to say that the U.S. and USSR were engaged in an arms race. Throughout the 1970s, as Albert Wohlstetter showed in *Foreign Affairs* magazine, the United States had actually been slowing its military spending while the USSR had been rapidly increasing its investment. As Patrick Glynn wrote in *Closing Pandora's Box*:

> Beginning in 1968 ... U.S. military spending began a long and steady decline, while Soviet military spending during the same period continued to increase. Essentially, a lessening of effort on the U.S. side did not result in the predicted reciprocal restraint from the Soviets. On the contrary, the greatest surge in the Soviet strategic buildup came *after* the signing of SALT I, when as part of a massive modernization effort, the Soviets deployed four new ICBMs, three new SLBMs, five new ballistic-missile submarines, and a new medium-range bomber.[25]

Testifying on Capitol Hill, President Carter's secretary of defense, Harold Brown, explained it in simple yet memorable terms: "We have found that when we build weapons, they build. When we stop, they nevertheless continue to build."[26]

But Brown's comments, though echoed again and again in conservative journals and newspaper columns, fell into a black hole on the Democratic side of the aisle. The liberals were unable to process Brown's report because they were too deeply in the grip of "mirror imaging." They assumed that the Soviets were motivated by exactly the same considerations as a Democrat from Massachusetts or New York. And since the only reason a Democrat could imagine voting for defense spending was to match a Soviet expenditure (if then), they naturally assumed that the Soviets were always responding to us. When it was pointed out that the Soviets were sprinting ahead in some areas, liberals would attribute this to "paranoia," never to belligerence. Because liberals told themselves that the Soviets were only responding to our military build-ups, it was no wonder that they regarded the "arms race" as a folly.

In his memoir, *In Confidence*, former Soviet ambassador to the United States Anatoly Dobrynin confirmed what many anticommunists had been saying for decades but was hotly disputed by liberals—that Brezhnev conceived of detente, not as a way to ease American-Soviet tensions or make the world safer, but simply as part of the "Marxist-Leninist 'class approach' to foreign affairs, which cast even peaceful relations into a mold of confrontation."[27] Though liberals always justified or explained the Soviet arms buildup as evidence of Soviet "fear of encirclement" or "insecurity," Dobrynin leaves no doubt that the Soviet plan was to "catch up with America and leave it behind."[28]

Certainly one did not need to wait for Dobrynin's memoir to divine Soviet intentions. But since arms control was sacrosanct to liberals, evidence was dismissed. Condemning the "military-industrial complex" was a constant refrain, and many Democratic members of Congress found fault with every weapons system for which they were asked to vote. Pointing to some future system that would theoretically work better, many would vote against nearly all military expenditures. In 1984, 194 Democratic members of the House of Representatives voted to bar funding of the MX missile. In 1987, 195 voted to forbid testing on a space based "kinetic-kill vehicle." Two hundred nineteen Democrats voted to urge the president to maintain the unratified SALT II limits without regard to Soviet compliance. A smaller number, but still a clear majority of Democrats (134 of them), voted against developing the neutron bomb. In 1986, while the Cold War was very much a going concern, seventy-six Democrats voted for a resolution proposed by Colorado Democrat Patricia Schroeder to cut U.S. troops devoted to NATO by 50 percent over five years. One hundred forty-five Democrats voted for an amendment to the defense authorization bill proposed by California Democrat Ron Dellums to bar funding for the B-1 bomber. And 146 voted to prevent modification of submarines to carry Trident II missiles.[29]

As the *Washington Post* acknowledged in a 1988 editorial, "A sizable body of Democrats would vote at the slightest jerk of a knee to roll back the military budget or freeze it—which means to let inflation roll it back untouched by human hands."[30]

The Democrats in Congress were contemptuous of Reagan's argument that peace could be best secured through strength. Michigan Democrat George Crockett dismissed the notion as fantasy:

> It's an *Alice in Wonderland* concept that you build more weapons to promote peace. If that were truly the case, the billions of dollars of military equipment and training the U.S. exports every year would have brought peace to Lebanon, and to El Salvador, and to Northern Ireland, and to South Africa. Obviously, weapons aren't the answer to the world's problems.[31]

When the Reagan administration put forward the MX missile to modernize America's land-based ICBM force, it was met with impassioned opposition from liberal Democrats in Congress. Massachusetts Democrat Ed Markey disdained the proposal:

> [According to the president, the MX missile] vote will be on sending signals to our NATO allies, on impressing the Russians with a bargaining chip and on testing our national will.... Yes, the MX is a bargaining chip. It is a bargaining chip with Congress to get Ronald Reagan to be serious about arms control. Yes, the MX is a test of our national will, but not against the Russians. This is a test of our will to stand up to the White House and the Air Force lobby. Mr. Speaker, the President says that he will accept a builddown for the MX. Fine. I have my own builddown proposal: For every MX deployed, two administration officials who supported the MX will clean out their desks and quit. Then we will see this country move ahead rapidly with a rational arms-control policy.[32]

Geraldine Ferraro, the Democratic candidate for vice president in 1984, said in her debate with George H. W. Bush:

Let me say first of all that I don't think there is any issue that is more important in this campaign, in this election, than the issue of war and peace. And since today is Eleanor Roosevelt's one hundredth birthday, let me quote her. She said, "It is not enough to want peace, you must believe in it. And it is not enough to believe in it, you must work for it." This Administration's policies have indicated quite the opposite.[33]

Her running mate, former vice president Walter Mondale, sounded similar themes. "The fact is," Mondale declared, " that four years of Ronald Reagan has made this world more dangerous. Four more will take us closer to the brink."[34] On another occasion Mondale said:

To put it bluntly, I believe that Mr. Reagan operates from fundamentally flawed premises about preventing war and keeping peace....In the nuclear age, the perception of strength is part of our arsenal. Yet Mr. Reagan persists in telling the world we're weak. Every time he unveils a new chart to prove American military inferiority, he undermines our confidence, frightens our friends, and tempts our adversaries. Every time he says we're weak, he literally weakens us.[35]

This is an extraordinary statement, implying as it does that the Russians would gauge our strength or weakness purely on the basis of rhetoric. The Soviets built their military capacity because it was noticed by other nations. And they certainly took note of our failure to augment our own defenses during most of the Carter years. They had spies, satellites, radar, and surveillance aircraft just as we did. They knew, possibly better than Vice President Mondale did, what the state of our military was. But since Mondale was opposed to nearly every weapons expenditure Reagan proposed, his policy would seem to amount to: "Be weak, but just don't talk about it."

Wisconsin Democrat Les AuCoin gave voice to the bitterness Reagan evoked in liberal congressional Democrats:

> Whether we are supposed to be bargaining with the Russians or with the Reagan administration is not clear. [Democratic Representative Les] Aspin says buying thirty MX missiles this year would give away the store to the administration, buying zero would give it away to the Soviets, so let's buy fifteen. It won't work. *We can't bargain with Reagan* [emphasis added].[36]

Madeleine Albright, who advised Walter Mondale on foreign policy in the 1984 campaign, scolded, "We have a president who seems to have a mind-set against arms control."[37] Though she would later claim, after becoming secretary of state in the Clinton administration, that her worldview had been forged by "Munich," Albright sounded like a conventional liberal Democrat in the 1980s. She scorned "Reagan's definition of strength ... destabilizing weapons like the MX [liberals always characterized American weapons as "destabilizing," but never described Soviet weapons that way], an unnecessary bomber like the B-1, and an unrealistic defense system like Star Wars."[38] In 1984, the Soviet Union chose to boycott the Olympics being held in Los Angeles. This was in retaliation for the American boycott of the 1980 games after the USSR invaded Afghanistan. With the economy in recovery, most Americans took the Soviet and East bloc decision in stride. For die-hard believers in arms control and detente, however, the absence of the Soviets was a terrible blow. Senator Gary Hart said:

> I think it's unfortunate and tragic. The Reagan administration has to understand that our relationships with the Soviet Union spring from whether or not we're achieving arms control. If we're not achieving arms control, then it spills over into and colors every other aspect of our relationship. And I can't help but believe the decision not to

attend the Olympics had a lot to do with the fact that this
nation was not making more progress in the direction of
reducing the arms race.[39]

Our relations with the Soviets were not, of course, based only upon arms control. They were affected by trade, restrictions on dual use technology, grain sales, regional conflicts, measures like the Jackson-Vanik amendment that linked "most favored nation" trade status to free emigration, our relationship with China, tensions over the Middle East, and human rights, to name just a few. But it is illustrative of the liberal point of view that Hart believed arms control to be the alpha and omega of the relationship. It was not containment, or deterrence, or the spread of democracy and free markets. Hart spoke the truth: For many liberals, their only goal was arms control.

## THE CHURCHES WEIGH IN

Liberal members of Congress were not alone in their antipathy to nuclear weapons and American military strength. In the early 1980s, they received key moral support from a "pastoral letter" prepared by the National Conference of Catholic Bishops. Rightly or wrongly (and the impression was mostly out of date), the Catholic Church was regarded as a conservative force in American life. The pastoral letter therefore gave an enormous boost to the unilateral disarmers. The letter put the Church's imprimatur on the then fashionable idea being propagated most notably by former defense secretary Robert McNamara, former national security advisor McGeorge Bundy, and containment father George F. Kennan, that the United States should adopt a "no first use" policy on nuclear weapons.

"No first use" was a deceptively attractive idea to populations grown fearful of a catastrophic nuclear exchange. Each side would pledge not to launch a dreaded first strike on the other. It seemed only reasonable. And yet, in the subtle world of deterrence and balance of power, it was not. In fact, the United States had implicitly promised, through its NATO agreement, to use nuclear weapons against the Soviet Union if that country ever attempted to conquer the nations of Western Europe. Our nuclear weapons were the counterweight to the Soviet Union's lopsided advantage in conventional forces in Europe. That is why the USSR had cheerfully assented to a "no first use" policy. The United States could not—not without potentially dire consequences.

The bishops went further, questioning the morality of nuclear deterrence itself, and adopting the view, based on Catholic "just war" theory, that use of nuclear weapons could not meet the criteria, namely, "probability of success," "proportionality," and "discrimination." John Cardinal Krol, testifying on behalf of SALT II, explained that deterrence itself was immoral:

> The moral judgment of this statement is that not only the use of strategic nuclear weapons, but also the declared intent to use them involved in our deterrence policy is wrong. This explains the Catholic dissatisfaction with nuclear deterrence and the urgency of the Catholic demand that the arms race be reversed.[40]

Though it may caricature the bishops' position a bit, and while it is clear that careful thought went into the pastoral letter, it is difficult to see how its reasoning differs from the "better red than dead" logic of the disarmament movement. Cardinal Krol, for example, was prepared to accept that the United States might be militarily defeated if it adopted his prescriptions:

> This does not mean we accept as inevitable the conquest of the world by a totalitarian system.... History goes on and political systems are subject to change. As long as life exists there is hope, hope that God's grace will enable suffering and oppressed peoples to endure.[41]

Not exactly the Church militant. But Cardinal Krol hardly marked the leftward boundary of ecclesiastical thinking. Archbishop Hunthausen of Seattle urged Catholics in his diocese to withhold 50 percent of their taxes on the grounds that this represented the share that was devoted to military expenditures. Bishop Matthiesen of Amarillo recommended that loyal Catholics give up their jobs in nuclear arms plants. And Father Francis X. Winters, a professor at Georgetown University, declared:

> Our security is not compatible...with the use of nuclear weapons. It is compatible, though arduously so, with military defeat.... Security depends on restraint, the unilateral renunciation of the intention to light the fuse that links the superpowers. Security may dictate surrender.[42]

A dismayed Michael Novak, writing in *Commentary*, inserted some common sense:

> Throughout the postwar era, it is the American deterrent that has kept the nuclear peace. Renouncing that deterrent would be as sure a way of bringing about war as one could devise. Such a consequence cannot be moral.... It is not "better to be red than dead;" it is better to be neither. As the history of our time amply demonstrates, some choosing the latter have not avoided the former. Avoidance of both sickening alternatives is the moral good which deterrence, and deterrence alone, effects. [43]

Nor were the Catholics the only religious group in America to agitate for disarmament.

The Protestant churches, at least the mainline ones, all hewed to the antinuclear line. Most had done so for quite some time. The Federal Council of Churches, precursor to the National Council of Churches, had issued a report in 1946 that reflected a profound antinuclear feeling:

> We would begin with an act of contrition. As American Christians, we are deeply penitent for the irresponsible use already of the atomic bomb. We are agreed that, whatever be one's judgment of the ethics of war in principle, the surprise bombings of Hiroshima and Nagasaki are morally indefensible.... We have sinned grievously against the laws of God and the people of Japan. Without seeking to apportion blame among individuals, we are compelled to judge our chosen course inexcusable.[44]

Later the National Council of Churches would endorse SALT II (which even a Democrat-dominated Senate would never ratify) as well as the nuclear freeze. The House of Bishops of the Episcopal Church condemned "massive nuclear overkill." And both the American Lutheran Church and the Lutheran Church in America endorsed the "no first use" policy.[45] Bishop James Armstrong, president of the National Council of Churches, announced that he found it "obscene" that the U.S. had decided to name a new nuclear submarine the *Corpus Christi*. "The name means 'body of Christ,'" he objected. "Jesus Christ stands in direct opposition to everything nuclear weapons represent."[46] The American Friends Service Committee, an arm of the traditionally pacifist Quaker church, agitated for unilateral disarmament.

The Reverend William Sloane Coffin, veteran of a thousand radical causes, advised that "Christians have to say that it is a sin not only to use, not only to threaten to use, but merely to build a nuclear weapon."[47] Coffin was a font of quotable one-liners on disarmament. "When I think of the ever escalating arms race," he once told a newspaper, "I think of alcoholics who know that liquor is deadly and who, nevertheless, can always find one more reason for one more drink."[48] The American Lutheran Church expressed concern over "the increasing sense of insecurity and peril to which our world is being led by escalation in nuclear weaponry. We see that our nation is locked with the Soviet Union in an arms race which both countries find almost impossible to stop."[49] And the American Baptist Churches issued a statement claiming that, "The presence of nuclear weapons, and the willingness to use them, is a direct affront to our Christian beliefs and commitment."[50]

The Rabbinical Assembly of America, representing conservative rabbis, endorsed the nuclear freeze. And Rabbi Alexander Schindler, president of the Union of American Hebrew Congregations (a Reform group) inveighed against the "most frightening" aspect of Ronald Reagan's presidency, "the escalation of the nuclear arms race by word and deed."[51]

"There is a movement starting again around the issues of disarmament and the poor," proclaimed Joan Campbell of the National Council of Churches. "It is similar to the feeling in the early 1960s around civil rights."[52] Ms. Campbell was always in the forefront of the National Council's leftist agenda, reveling in its support for the PLO and for Fidel Castro.[53]

## THE ESTABLISHMENT LEFT

The movement for disarmament did not engage the conscience of the nation as the civil rights movement had, but it did win the enthusiastic participation of a grab bag of old leftists, Communists, ad hoc groups, aging hippies, earnest housewives, young professionals, as well as a good portion of the liberal opinion makers in the nation. Groups sprang up with names like Ground Zero, the Council for a Nuclear Weapons Freeze, Physicians for Social Responsibility, the Union of Concerned Scientists, Women's Strike for Peace, Mobilization for Survival, the Women's International League for Peace and Freedom (actually begun by Jane Addams), the Committee for Peace in a Nuclear Age, and many more. Additionally, the movement's numbers were swelled by the participation of unions, church groups, the Americans for Democratic Action, the Citizens Party, the Communist Party USA, and others.

They received support from major American foundations like the Rockefeller Family Fund, the Stern Fund, and the Ploughshares Fund. An early supporter of antinuclear activities was the small New World Foundation, which had allocated more than a quarter of a million dollars for "research and initiatives to broaden grass roots participation" in antinuclear agitation. One of the group's members was Hillary Rodham Clinton.[54]

There were occasional dissents on the Democratic side of the aisle, but they were rare. Senator Henry Jackson of Washington State, as was his wont, inserted a bit of ballast to counterbalance all of this hot air. In a 1981 meeting with President Reagan, he noted the success the Soviet "peace" initiative was having in Europe and urged the president to launch a "dramatic, sustained American peace offensive."

> It will clarify to our friends and allies our dedication to stabilizing nuclear arms reduction negotiations. At a minimum, it will put Moscow on the defensive. It keeps open the possibility of achieving over time real and significant steps in the direction of such a stabilizing disarmament at a time when the Soviet economy is in deep difficulty.[55]

In the event, it was a former Jackson staffer, Assistant Secretary of Defense Richard Perle, who originated the idea Reagan would introduce to diffuse the Soviet "peace" offensive—the "zero option." If the Soviets would withdraw their intermediate range missiles from Europe, NATO would not deploy any. It was simple, elegant, and reminded the world that the Soviets were the aggressors in this confrontation.

There was one other Democrat who cautioned his party against its tilt toward appeasement. Wisconsin congressman Les Aspin became disgusted with his fellow Democrats after viewing a debate among presidential aspirants in 1984. They were "falling all over each other in trying to be more antinuclear than the other," Aspin lamented. "The perception is that they are denouncing nuclear weapons. [It was more than perception.] It's a debate that leads you to believe that they're way off on the edge." Aspin, who voted in favor of the MX missile, was nonetheless critical of many aspects of Reagan's foreign policy. But, he said, "It ought to be possible to stake out a position to the left of Ronald Reagan and not be crazy about it."[56] It was a maneuver that most Democrats found too difficult to manage.

"The problem facing our nation and the world is that Mr. Reagan's policies have contributed to an increasingly dangerous arms race," declared Walter Mondale during that debate. He went on:

> Nuclear arms control is *the* issue of our time. There is none that compares with it.
>
> And one of the things I would like to do is to institutionalize annual summit conferences where we would meet at least once a year with the head of the Soviet Union. We know all of our differences, but to sit down and see if we can't in the name of humanity put some sense in this relationship, freeze those weapons, reduce their risks, and bring some sanity to this ever increasingly dangerous world.[57]

Jesse Jackson offered that "the leadership must take the risk for peace and make a difference. If you meet with Andropov, if you meet, if you talk, you act: if you act, you change things. So you must—you must unthaw that relationship before you can then agree to freeze verifiably the weapons." California Democratic senator Alan Cranston, who based his entire candidacy on the nuclear freeze, was more specific: "If I am elected, on January 20, 1985, the day I take the oath of office, I will announce that the United States will not test or deploy any more nuclear weapons as long as the Soviets do not test or deploy. The next step would be that I would, on that same day, get in touch with the Soviet leader, Yuri Andropov or whoever is then in charge there, and suggest a very early meeting between myself and that leader and our top advisers to discuss further steps."[58]

The zero option made a difference among some fair-minded people. But most actors and artists did not fall into that category. Artist Robert Morris did a drawing, "Firestorm," which was displayed in the mezzanine of the Museum of Modern Art in New York. Harry Belafonte leant his prestige to the disarmament campaign, along with Colleen Dewhurst, Jules Feiffer, Meryl Streep, Kris Kristofferson, and Robert Blake, along with many others.[59] Martin Sheen went to the trouble to get himself arrested trespassing on a nuclear test site in Nevada.[60] The Reverend Jesse Jackson participated in a "Peace Sunday" in Los Angeles, along with Jane Fonda, Ed Asner, Muhammad Ali, and others.[61]

Certainly some of the impetus for the sudden pollination of all these groups was Reagan's new tough rhetoric, and occasionally, even his humor. Preparing for a weekly radio address, unaware that his microphone was open, the president once joked: "My fellow Americans, I'm pleased to announce that I have signed legislation that would outlaw Russia forever. We begin bombing in five minutes."[62] Critics were surely right that nuclear war was too serious a matter to joke about in front of a mike, even if you believed it to be off. But the response of earnest arms controllers and liberals everywhere (in the U.S. and abroad) was nearly hysterical, showing that they feared Reagan perhaps even more than the Soviets did. And certainly they feared Reagan far more than they feared the Soviet Union.

This fear was unleashed again when President Reagan said, in an interview with newspaper editors, "You could have an exchange of tactical weapons against troops in the field without it bringing either one of the major powers to pushing the button." This caused something close to panic among the disarmament groups. But the comment was worrisome only if you assumed that Reagan was itching for a land war in Europe and thought we could win it without resorting to a full nuclear exchange.

But under what scenario—other than a Soviet attack against Western Europe—would there be a battlefield exchange of tactical nuclear weapons? There was simply no circumstance in which it was conceivable that the United States would launch an unprovoked attack against the USSR. So Reagan was merely musing about the possibility that Europe might be defended without U.S. and Russian cities being obliterated.

That this caused unease among the Europeans is understandable, as one possible interpretation of Reagan's remarks was that the United States was imagining a conflict that would take place entirely on European soil—thus "Europeanizing" the Cold War. But the willingness to use nuclear weapons in self-defense was a pillar of the Western Alliance, and the fact that mere mention of such a possibility caused sizable numbers of liberals and leftists to get attacks of the vapors in the early 1980s was further evidence of their complete demoralization in the face of the Soviet threat. They had, in short, a total loss of will for the confrontation with Communism.

## The Fear Mongers

While Reagan's rhetoric may have catalyzed a new activism on the Left, there is little doubt that it also inspired and encouraged those in the center and on the Right, who had never believed that their country was the dangerous or destructive force it was painted as by the Left. As noted above, Reagan's solidarity with the oppressed peoples of the Communist world offered hope to millions behind the Iron Curtain. And in the West as well, despite the hundreds of thousands who marched and protested, millions more were persuaded that Reagan's view of the conflict between East and West was correct.

Still, Reagan was up against an enormous headwind. In 1982, Jonathan Schell published *The Fate of the Earth*, a series of essays that had run in the *New Yorker* magazine about what would happen in the event of a nuclear war. The book was graphic, detailed, and frightening, and it became a bestseller. The CBS *Evening News* gave the book favorable mention, as did Helen Caldicott ("the new Bible of our time"). Bill Moyers mused about what would happen if Ronald Reagan and Leonid Brezhnev read the book to one another at the next summit meeting, while Walter Mondale declared the book to be "historic."[63] The premises underlying Schell's predictions were simply new iterations of hoary arms-control myths. One was the notion that as stockpiles of weapons multiply, so do the chances of accidental nuclear war. But both the Soviet Union and the U.S. had taken steps by the early 1980s to make the risk of accidental launch much lower than it had been a decade earlier.

There was in Schell's work and that of other disarmament advocates an absence of moral clarity about the nature of the Communist threat. No mere political value, not even freedom, seemed to them to be worth risking nuclear war. Not every liberal was moved to genuflection by Schell's work. *Harper's* editor Michael Kinsley declined to join in the general adulation. He called *The Fate of the Earth* "one of the most pretentious things I've ever read." Schell's writing amounted to little more than "bullying," Kinsley wrote, and featured "hothouse reasoning: huge and exotic blossoms of ratiocination that could grow only in an environment protected from the slightest chill of common sense."[64]

But that fairly well describes the environment of the United States in the 1980s. It was an atmosphere composed of equal parts dread and hokum—a childlike perspective. Indeed, the innocence of children who want only world peace was uncritically embraced and celebrated by adults who should have known better. Certainly any adult with common sense should understand that the mere wish for peace, untethered to realistic self-defense, is not a safe posture in this world, and further, that signaling weakness to a bully is probably the most dangerous course possible. But a great many adults made fools of themselves over "space bridges" to the Soviet Union, "Citizens Summits," "sister city" initiatives, and so forth. Some communities went so far as to declare themselves "nuclear free zones," whatever that meant. TV performer Phil Donahue probably earned the title "useful idiot" as much as anyone during the 1980s, as he credulously accepted the apologetics of his favorite Soviet "journalist," Vladimir Pozner, and provided a forum for every left-wing cause imaginable on his popular television program *Donahue*. In 1987, he happily accepted a joint award with Pozner from the Better World Society, a Ted Turner organ that also supported the Sandinistas in Nicaragua.[65]

With Pozner, Donahue hosted a series of programs linking Soviet and American audiences. And while Pozner served as an indefatigable apologist for the Soviet Union, Donahue maintained a fine impartiality. "There are 60,000 nuclear bombs on both sides," Donahue wailed, "When is this madness going to stop?" On other occasions, Donahue went out of his way to tell members of his Soviet audience that while "some Americans worry when the Soviet Union expands . . . the vast majority of people in the United States admire you." Donahue may have admired the USSR; there is little evidence that a "vast majority" of Americans shared his views. Donahue was moral equivalence personified. The trouble between the superpowers, he explained, was the fault of "a small percentage of people in both countries—yours and mine—who remain hard-line and militaristic."[66]

Vladimir Pozner was all the rage on American television during the 1980s. Born in the U.S., he lived in Brooklyn until his teen years when his Communist parents moved first to East Germany and then to the Soviet Union. His unaccented English distinguished him from the typical Soviet spokesmen of earlier times. But his undeviating defense of the regime as well as his lies and distortions marked him as utterly in the mold of his predecessors. He claimed, for example, that Soviet authorities failed to explain the Chernobyl disaster for several days due to lack of information, when in fact, they had already dispatched doctors and emergency workers to the scene. Regarding the Korean Air Lines Flight 007, Pozner maintained to the bitter end that it was a spy plane.

But Pozner was extremely popular with American television producers. He became a regular on ABC's *Nightline*, made dozens of appearances on *Donahue*, and was interviewed for countless news programs. ABC even gave him eight minutes of uninterrupted airtime following one of President Reagan's speeches. (The network apologized for this after receiving a tart complaint from the White House.) When the magazine *MediaWatch* examined the way Pozner was identified, they found that he was referred to as a "Communist" only once out of 157 references. Most called him simply a "commentator," which was misleading since it implied objectivity. Twenty-one stories described him as a "spokesman," which was a lot closer to the truth. And thirteen described him—utterly falsely—as a "journalist."[67]

The same credulity that permitted Pozner such a long run on the American stage elevated little Samantha Smith, a Maine fourth grader, into an international heroine. Smith got rapturous attention for a letter she penned to Yuri Andropov, expressing her worry about nuclear war. "I have been worrying about Russia and the United States getting into a nuclear war," she wrote, in a letter that *Pravda* reprinted. "...Please tell me how you are going to help to not have a war." Andropov did not do that, but he did invite Smith and her family for an all expenses paid trip to the Soviet Union and generally made propaganda hay of her. Manchester, Maine, gave her a parade upon her return. She sat perched on the rear seat of a convertible, roses in her lap, and waved to the cheering crowd like the homecoming queen she was. The American press made her a young heroine, and there was lots of talk about the "wisdom" of the young. Columnist Ellen Goodman wrote:

> Samantha was just a kid who thought like a kid. She woke up one morning in Maine when she was 10 years old and "wondered if this was going to be the last day of the Earth." She read about the arms race and thought "It all seems so dumb to me."
>
> And then she did something that only an unsophisticated kid might do, in the years before diplomacy breeds

directness out of them, before cynicism and a sense of pow-
erlessness sets in. She wrote a letter to the Soviet leader....

When Samantha came home, she wrote..."I mean, if
we could be friends by just getting to know each other better, then what are our countries really arguing about?
Nothing could be more important than not having a war if
a war would kill everything."

Adults cannot say such things anymore. Adults must
talk about SALT and START treaties, about Star Wars this
and MX that, about parity and verification. Adults must
be suspicious, cautious.... So we ask children to express
the fears that we share and the idealism that is, finally, our
hope.[68]

But of course, adults were certainly allowed to say such things and
were saying them ad nauseum. As for the childish sense of directness,
it is not always supplanted by "cynicism and a sense of powerless-
ness." Some adults simply lose their naiveté and face life's harsher
realities, including the reality that it is sometimes necessary to protect
oneself with weapons against those who have evil intentions. It was
this maturity and realism that was hard for many American wishful
thinkers to acquire. In fact, they disdained steady judgment and maturity, preferring to call self-delusion and childishness "idealism."

Samantha Smith could have been the spokesman for the Democratic Party on the subject of the conflict between the superpowers.
It, too, had a weakness for the "if we could only just be friends" style
of analysis. It, too, believed that "nothing could be more important
than not having a war." And it, too, failed to see why we couldn't all
get along by seeking to know one another better. It was a trope of liberal analysis during this period (and similar arguments are now
advanced vis-à-vis other adversaries like the Islamists) that conflicts
arise among nations due to "misunderstandings." The cure for these
misunderstandings could be found in summit meetings, cultural

exchanges, sister city initiatives, pen pal programs, and "space bridges." Because the diagnosis was flawed, the cures were unsuccessful. As it turned out, those who understood the Soviets best were the very "hardliners" so scorned by liberals—a fact that was testified to by many former Soviets after 1991. It wasn't that both nations, essentially benevolent, mistrusted one another due to the accumulation of weapons on both sides in some sort of mad and inexplicable race to destruction. Rather, the Soviets were aggressive and predatory, and the United States (along with its allies) sought to thwart it. When the Moscow regime changed, the so-called "arms race" was over.

Doubtless Samantha Smith's letter to the leader of the USSR was motivated in part by the beliefs of her parents. Her mother, Jane Smith, told reporters that Samantha "thinks it would be better to spend more money on programs for the poor rather than on bombs."[69] She was even asked to question the Democratic candidates for president. (Smith died tragically in a plane crash a couple of years later.) But she had many imitators. The "teachers of peace," ten school children from California made a similar pilgrimage to Moscow a couple of years later. The National Council of Churches organized a "crayon brigade" of children who wrote to their counterparts in the USSR expressing friendship and best wishes.[70] Just as Jimmy Carter had taken advice from daughter Amy about nuclear proliferation, the nation was to take the counsel of elementary school children on world peace.

Innocence is a precious quality in children. It is less appealing in adults. It is particularly unappealing when it is put in the service, whether intentionally or not, of a deeply cynical and criminal enterprise like the Soviet Union.

Samantha Smith had one more imitator, but this one did not get worldwide attention and acclaim. Irina Tarnopolsky, age twelve, heard about Andropov's correspondence with the American girl, and was inspired to write a letter of her own to the Soviet leader. She wrote to Yuri Andropov pleading for him to release her father from prison. Yuri Tarnopolsky, a research chemist who taught at Krasno-yarsk, had applied for an exit visa for himself, his wife, and their daughter. Permission was refused and Tarnopolsky was dismissed from his position as a professor. Tarnopolsky then joined other Soviet Jews in agitating for freedom of emigration. He also smuggled out a book of poetry which was published in France. At this, he was arrested and sentenced to three years in a labor camp for "slandering the Soviet system." He was also accused of "parasitism" by the KGB—for being out of work.[71]

Irina never became an international celebrity.

## WEAPONS AND THE MINDS BEHIND THEM

The idea that weapons cause wars has a long pedigree. It began as spin for Imperial Germany. In his excellent history of arms control, *Closing Pandora's Box*, Patrick Glynn traces the history of what he calls the "Sarajevo fallacy." This fallacy is familiar. It is the widely accepted version of how World War I began. The great nations of Europe had engaged in an uncontrolled arms race in the ten years before 1914. They were so bristling with armaments, goes the tradi-tional view, that when Archduke Franz Ferdinand was assassinated in Sarajevo, it served as the spark to ignite all those weapons into a gen-eral conflagration.

That is not how the origin of the war was seen at the time. Cer-tainly among the allies, guilt was laid at Germany's door. As Glynn shows, the allies were right, but a "revisionist" effort succeeded in dis-torting that history:

> Eager to free itself of the charge that it was responsible for the war—a charge that legitimated the onerous reparations demanded in the Versailles Treaty—Germany mounted a massive propaganda campaign involving extensive (though selective) publications of state documents. In an effort to

discredit the overthrown Russian monarchy, the newly installed Bolshevik regime in Moscow likewise began to release selected archival publications. These documents seemed to shift blame away from Germany.... Under these influences, historians, especially in America and Britain, increasingly began to play down German responsibility, shifting blame more and more to impersonal factors such as militarism, nationalism, and the arms race itself. This view in turn became deeply ingrained in Western thinking about war and peace.[72]

But as Glynn carefully documents, the image of an arms race between Imperial Germany and Great Britain in the years before World War I (a view popularized by historian Barbara Tuchman in her Pulitzer Prize winning book *The Guns of August*—a book that John F. Kennedy insisted his entire foreign policy staff read) was quite oversimplified. There was a rivalry between Germany and Britain. But the British government contained a powerful pacifist wing that succeeded in slowing British production of large battleships, destroyers, and submarines. "We desire to stop this rivalry," the prime minister, Sir Henry Campbell-Bannerman, told the House of Commons, "and to set an example in stopping it."[73] Great Britain also pushed hard for disarmament talks at The Hague. The German response, however, was to redouble its military efforts. The Foreign Office senior clerk would write in 1907: "Our disarmament crusade has been the best advertisement of the German Navy League and every German has now been persuaded that England is exhausted, has reached the end of her tether, and must speedily collapse, if the pressure is kept up."[74]

Just as American policymakers would engage in "mirror imaging" when it came to interpreting the behavior of their Soviet adversaries, British statesmen made the identical error vis-à-vis the Germans prior to World War I. With a surprising failure of imagination, they tended to assume that Germany's motivations and goals were similar to Great

Britain's. The Germans engaged in mirror imaging as well. When they saw Britain failing to push its advantage, they concluded that she must be on her last legs. Only weakness, after all, would have induced the Germans themselves to act in such a fashion. Thus British goodwill was interpreted as weakness in Berlin, while German aggressiveness was interpreted as insecurity in London. At other times, the Germans saw quite clearly that Britain was acting on benevolent or at least pacific impulses, but used this as an opportunity to gain the upper hand. When it seemed that a Tory government might succeed the Liberal party in England, the German ambassador was quick to warn the Kaiser: "A conservative government...would represent a very real danger for us....We should do all in our power to keep the Liberal party, to which all peace-loving elements in England adhere, at the helm."[75]

The true cause of World War I, Glynn argues, was Imperial Germany's drive for European dominance. About sixty years later, President Jimmy Carter made the decision not to build the B-1 bomber. As Glynn recounts:

> Shortly after the Carter B-1 decision, Senator John Tower
> of the Senate Armed Services Committee was in Moscow.
> He asked Alexander Shchukin [a Soviet arms control nego-
> tiator] what the Soviets were prepared to do to reciprocate
> Carter's cancellation of the B-1. "Sir," responded Shchukin,
> "I am neither a pacifist nor a philanthropist."[76]

Another argument that unilateral disarmament advocates relied upon was that of "overkill." With so many warheads on both sides, they maintained, there was no point in debating which side had an advantage over the other (though they always vociferously denied that the Soviets had an actual advantage). Both superpowers, ran the argument, had the power to destroy one another "many times over," and so it was foolish and absurd to speak of "winning" a nuclear confrontation.

It may have been true, as the Reagan administration found itself constrained to say over and over again, that "a nuclear war can never be won and must never be fought." But it was hardly the case that mere ownership of a minimum number of nuclear weapons was sufficient to deter an adversary like the Soviet Union. If that were the case, then Europe would have had no need of NATO since Britain and France were both nuclear nations capable of inflicting damage on the USSR. The Soviet Union certainly seemed to think it was purchasing something by its enormous investment in upgrading its land based nuclear arsenal. And despite predictions from the likes of George Kennan that the Soviets had no further territorial ambitions, the possession of overwhelming military power carries with it diplomatic, political, and psychological advantages. The psychological impact of Soviet power was evident in the response of European and American liberals throughout the Cold War period. Certainly some part of the liberal willingness to appease and excuse the monstrous regime in Moscow was based on crude, cold fear. Many had had the same reaction to Hitler's Germany in the period before World War II. His blatant military buildup, domestic repression, and hate-inspired propaganda frightened some in the West and led them to counsel appeasement. In the aftermath of the war, the policy of appeasement was rightly regarded as cowardly and dishonorable. In the aftermath of the Cold War, no less can be said of those who counseled appeasing the Soviet Union.

Though Robert McNamara publicly assured the American people in 1965 that "there is no evidence that the Soviets are seeking to develop a strategic nuclear force as large as ours," that is exactly what the USSR was attempting to do. In fact, the Soviets achieved parity and then went further, striving for superiority. The U.S. permitted this to happen, believing that through a balance of terror, peace would be maintained. In the decade of the 1960s, the Soviets deployed five new ICBMs, one new SLBM, and four new models of ballistic missile submarine. Though America's overwhelming nuclear superiority had been key to the peaceful resolution of the Cuban Missile Crisis, by 1972, it had been erased. The Soviets had 1,510 ICBMs in 1972—five hundred more than the United States. And by the early 1980s, the Soviets had achieved the capacity to destroy 90 percent of America's Minuteman missiles in their silos.[77]

As the Soviet Union was surpassing the United States in weaponry during the 1970s, it was achieving significant political and military victories on the ground. South Vietnam, Cambodia, Laos, South Yemen, Nicaragua, Grenada, Angola, Ethiopia, Mozambique, and Afghanistan had all joined the Communist camp between 1974 and 1980. Little wonder that Leonid Brezhnev could state confidently in 1976, "The general crisis of capitalism continues to deepen. Events of the past few years are convincing confirmation of this."[78]

Just as the Soviets had surged ahead of the U.S. in nuclear weaponry, an idea swept through the Western world that would have cemented Soviet superiority. It was called the nuclear freeze. The nuclear freeze movement drew strength from diverse sources, and considering the power and reach of its advocates, it is remarkable that the campaign did not in the end succeed.

Advocates of the nuclear freeze posited that nuclear war would spell the end of human life on earth. This was, to say the least, a highly dubious idea, since the earth and human beings had managed to survive huge volcanic eruptions, ice ages, plagues, and other disasters. And it was certainly quite possible that while millions of Russians, Americans, and perhaps Europeans would die in a full nuclear exchange (immediately, or worse, over the course of the next several months and years from radiation sickness), the idea that the remainder of the planet would become uninhabitable was sheer speculation. Moreover, it was obviously conjecture designed to inhibit the West's self-defense, not the aggression of the Communist world.

But liberals were very, very frightened and made it their business to frighten everyone else. It is worth pausing for a moment to analyze this fear, because it is revealing. It is certainly the case that nuclear war was a frightening prospect, and no sane person would be cavalier about it. It is also certainly true that nuclear war held certain terrors that previous wars had not—namely the wholesale annihilation of entire cities, even scores of cities on two continents, along with lingering radiation that might make survival less than enviable. One key challenge of statesmanship in the second half of the twentieth century, therefore, was steady nerves.

But a wholesale rejection of defense, and particularly of nuclear weapons, though it traveled under the name of peace, was actually a policy of slow surrender.

Just as the baby boom generation seemed to believe it was the first to discover sex, many of its members also seemed to think they were the first to discover the horror of war. Without diminishing the devastation that nuclear war would bring, it is worth recalling that all major wars in human history were frightening and horrifying, particularly when being on the losing side meant your homes would be pillaged, your cities sacked, your children dying of disease and malnutrition, and your women raped. All wars have certainly meant death and crippling injuries for soldiers as well as widowhood and bereavement for those left behind. And yet sensible human beings throughout most of history have understood that the best way to ensure the peace is to prepare for war. Further, war, while undeniably terrible, is not always the worst evil to be feared. Being overrun is usually worse. Certainly most American liberals were quite relieved to see the United States enter World War II, the bloodiest conflict in world history. Why? Because something frightened them more than war itself—namely fascism. When the alternative to war was brownshirted Nazis issuing decrees for Des Moines and Denver, plunging into a bloody and painful war seemed well worth it.

The Nazis were in a sense the perfect enemy for America. Their expressed beliefs in a master race and world domination were so antithetical to the virtues admired by Americans that hating and opposing them came quite naturally (though whether the U.S. would have declared war on Germany if Germany had not first declared war on us is an open question historically).

But the Communist enemy was different. Communists paid lip service to liberty, and pretended to have achieved equality. Communists in the United States participated in the civil rights movement and seemed on the right side of the conflict in South Africa. Their propaganda stressed solidarity with the downtrodden of the Earth. And if their actual conduct differed very little from that of the Nazis, indeed, as Paul Johnson argues in *Modern Times*, the Nazis learned a few tricks from Lenin, those were technical details. Communism's halo might have been askew, but it did not appear to liberals that resisting Communism was worth risking nuclear war.

Why not just state this openly? Why didn't the liberals argue, as many European leftists did, "better red than dead." Because one could not make the case to a largely anticommunist America that fighting Communism was not worth the candle. Though the elites might have found something to admire in the grand Communist enterprise, most Americans, particularly those with religious convictions, emphatically did not. And while anti-anticommunism had taken hold as the fashion among intellectuals, the broad American public continued to take a dim view of Brezhnev and Co. Liberals accordingly recognized that in order to make headway, they must up the ante. And they pulled out all the stops. It wasn't just about risking war, they warned, but the entire fate of the planet was at stake.

Carl Sagan, an astronomer at Cornell University and a famous television popularizer of astronomy, teamed up with Stanford biologist Paul Ehrlich (who had a poor track record with predictions, having wrongly prophesied that a "population explosion" would lead to widespread famine by 1985)[79] to propose a theory called "nuclear winter." Even an exchange involving only 10 percent of the nuclear arsenals on both sides, they warned, would cause a huge dust cloud to ascend into the atmosphere, blocking out the sun for months. Temperatures would plunge to –23 degrees centigrade, and radioactivity would remain high. Poisonous fumes from burning cities would travel the globe, and the ozone layer would be depleted causing dangerous levels of ultraviolet radiation to reach the Earth's surface (though whether the dust clouds would provide a layer of protection against ultraviolet rays, Professor Sagan did not say).

Soviet scientists quickly chimed in saying that their independent research had predicted the same thing—only worse. "A Soviet research team has found that the recently issued 'nuclear winter' report by one hundred American and European scientists actually understated the extent of the climatic disaster that would be caused by nuclear war," the *Washington Post* reported.[80] The *Post* report contained no cautions about the lack of independence of Soviet scientists or the interests of the Soviet Union in promoting antinuclear panic in the West.

Columnists and pundits chimed in, bouncing off the Sagan-Ehrlich horror scenario to perform their own pirouettes of panic. Ellen Goodman wrote:

> Most of us have grown up under the threat of extinction. In the past we behaved much like children who entrust their anxieties to powerful adults in the belief that responsible grown-ups will take care of them.... What has changed is that the adults in charge today are not careful enough. Indeed, they are menacing. From the early bulletins of "winnable" nuclear wars, to the invasion of Grenada, the Reagan administration has pursued security through an aggressive posture.... Instead of talking the Soviets back from the brink, like madmen tied to each other on the roof of a building, our leaders have spent the last three years daring them closer to the edge. Instead of building down, we are building up. The only thing we have dismantled is meaningful disarmament talks.[81]

Dr. Helen Caldicott, a pediatrician who headed Physicians for Social Responsibility, injected her usual emotionalism: "We are thinking of our babies; there are no communist babies; there are no capitalist babies. A baby is a baby is a baby."[82] Caldicott's use of the preferred communist terminology—"communist" versus "capitalist" instead of communist versus free—was typical of the nuclear freeze movement, which tended to minimize and even ridicule the differences between the two worlds. Anyone who insisted that those differences were morally significant was dismissed as an enemy of "peace." This subtle matter of word choice fitted perfectly with Soviet propaganda of the time. Did this make the freeze movement a creature of the KGB? No. But did it advance the Kremlin's interests more than the free world's? Yes.

Leading citizens, including George F. Kennan, former CIA chief William Colby, and Thomas J. Watson Jr., former U.S. ambassador to the USSR and onetime CEO of IBM, endorsed the nuclear freeze idea.[83] A poll in the spring of 1982 found 72 percent of Americans endorsing the idea.[84] Several states and hundreds of cities passed nuclear freeze resolutions. The American Bar Association passed a resolution calling for "serious negotiations to end the nuclear arms race" and an end to "conduct and rhetoric that invite nuclear confrontation."[85] Speaker of the House Thomas P. O'Neill Jr., described the nuclear freeze movement, without exaggeration, as "one of the most remarkable political movements I have ever seen during my years in public service."[86]

In October of 1982, twenty-six groups would band together to form Citizens Against Nuclear War, with a combined membership of 18 million. Member organizations included the American Jewish Congress, the American Public Health Association, Friends of the Earth, the National Council of Negro Women, the Newspaper Guild, the National Education Association, and the United Food and Commercial Workers International Union.[87]

Asked at a press conference whether there was evidence of Soviet infiltration of peace movements, President Reagan said yes. While the "overwhelming majority" of those who supported a nuclear freeze were "sincere and well-intentioned," the president explained, there was "plenty of evidence" that the Soviets had attempted to gain influence in these groups for their own purposes.

Within a nanosecond, the American Civil Liberties Union accused the president of McCarthyism. "It is disheartening," said Morton H. Halperin, speaking for the ACLU, "to see an American president returning to the tactics of McCarthyism. Charges of secret manipulation by foreign agents poison the well of public debate."[88] Or could it perhaps have been the business of antinuclear activists to police their ranks more carefully? Imagine if a Nazi group had teamed up with the Christian Coalition to protest something and a U.S. president mentioned the Nazi role. Would this be called McCarthyism? Never. The Christian Coalition would be swiftly vilified and forced to denounce and renounce any association. As for the truth of the Soviet involvement, there is evidence supplied by Soviet defectors as well as testimony from the assistant director of the FBI who told the U.S. Congress that Mr. Reagan's description of Soviet attempts to influence the nuclear freeze movement were "accurate."[89]

Serious panic took hold, bolstered by leading members of the Democratic Party. Democrat senator Edward M. Kennedy of Massachusetts and Republican senator Mark O. Hatfield of Oregon introduced legislation in 1982 calling for a nuclear freeze. "The arms race rushes ahead toward nuclear confrontation," Kennedy warned, "that could well mean the annihilation of the human race."[90] Tennessee Democrat Al Gore also introduced a slightly different freeze resolution when he served in the House.[91] The Kennedy measure had seventeen co-sponsors in the Senate and 122 in the House. Thanks to the intercession of Senator Henry L. (Scoop) Jackson, the Washington state Democrat, an alternative freeze resolution was introduced that would have required a freeze at equal and reduced levels of weapons. The existence of this alternative doomed Kennedy's freeze proposal, though it was close. A 1982 freeze proposal was defeated in the House by a vote of 204 to 202.[92]

Adding to the climate of fear and appeasement was a public television production called *Testament*, starring Jane Alexander (later director of the National Endowment for the Arts), a moving film about a family in Washington State slowly dying of radiation poisoning after a nuclear war. Even more sensational was the ABC television drama called *The Day After*. Starring Jason Robards, the film, which depicted the effects of a nuclear war on the city of Lawrence, Kansas, became a national event. Simplistic, heavily emotional, and tendentious, the movie was considered so important that Secretary of State George Shultz agreed to appear on a special *Nightline* program after the broadcast to discuss the implications. The panel also included Carl Sagan, Henry Kissinger, William F. Buckley Jr., Brent Scowcroft, Elie Weisel, and Robert McNamara. Commercial time during the broadcast was purchased by Democratic senator Alan Cranston from California whose presidential campaign in the Democratic primary rested heavily on a nuclear freeze platform and also featured actor Paul Newman instructing viewers how to write or phone to receive a "nuclear war prevention kit."

When questioning Secretary Shultz, Ted Koppel, an exquisitely tuned instrument of the conventional wisdom, asked, "With so many of them [nuclear weapons] is it not inevitable that at some point they will be used? And if not, why do we need them?" Schultz responded in measured tones that reduction in the levels of nuclear weapons was administration policy. Koppel was a bit impatient with this, demanding, "Isn't this business as usual in the most lamentable way—the Soviets are pointing fingers at us, and we are pointing fingers at them, and somehow the moral imperative of arms control is going nowhere. Why?"[93]

That is how it seemed to the conventional wisdom in 1983, the "moral imperative" was not to defend freedom or to contain Communism, but rather to push arms control.

NBC, not be outdone, broadcast its own scaremongering documentary called "Facing Up to the Bomb." It was advertised with promos like "The Nuclear Strategy Game: Are We Nearing Checkmate? Watch this Broadcast as if Your Life Depended on it. It may." Certainly NBC's revenues depended upon people watching. Perhaps

NBC was unconsciously or consciously imitating the so-called "Doomsday Clock"—a public relations gimmick by the Bulletin of Atomic Scientists. The Bulletin would measure the safety of the world by moving the hands of its Doomsday Clock closer to or further from midnight—which represented nuclear Armageddon. In 1983, the hands of the clock were moved from 11:48 to 11:57 after the Soviets walked out of arms control talks.[94]

## THE STRATEGIC DEFENSE INITIATIVE

But the Soviets returned the following year. What brought them back was Reagan's Strategic Defense Initiative, a weapons system that was almost universally derided as "Star Wars" by newspapers such as the *Los Angeles Times*, the *Washington Post*, and the *St. Louis Post-Dispatch*. Reagan proposed to invest in research on antiballistic missile systems that could destroy missiles before they hit the United States. As it was (and remains), the United States had absolutely no defense against ballistic missiles. The nation relies on deterrence alone. Reagan proposed that it would be safer and more humane to rely on defense instead of on the threat of massive retaliation. And he proposed to share the technology with the Soviets.

Though his proposal was widely ridiculed then and since, his argument on the day he proposed it was quite substantive. He acknowledged that:

> ...this is a formidable, technical task, one that may not be accomplished before the end of this century. Yet, current technology has attained a level of sophistication where it's reasonable for us to begin this effort. It will take years, probably decades of effort on many fronts. There will be failures and setbacks, just as there will be successes and breakthroughs.
>
> Defensive systems have limitations and raise certain problems and ambiguities. If paired with offensive systems,

they can be viewed as fostering an aggressive policy, and no one wants that. But with these considerations firmly in mind, I call upon the scientific community in our country, those who gave us nuclear weapons, to turn their great talents now to the cause of mankind and world peace, to give us the means of rendering these nuclear weapons impotent and obsolete.... We seek neither military superiority nor political advantage. Our only purpose—one all people share—is to search for ways to reduce the danger of nuclear war.[95]

The president's idea was laughed out of town. Columnist Molly Ivins scoffed: "Star Wars, you may recall, was the brainchild of Ronald Reagan, and in an effort to distance it from its origins in a couple of old movies that Reagan had seen, it was christened the Strategic Defense Initiative...."[96]

Whenever possible, the press would spin news to reflect badly on the SDI idea. After the failure of one interceptor test (the failure was in a secondary system with no bearing on the feasibility of SDI), the *New York Times*'s Michael R. Gordon predicted that the U.S. had "no choice...for years to come" but to stick with MAD.[97]

When an air force panel released a study on the feasibility of missile defense, the *Washington Post* headlined the story: "More Doubts Are Raised on Missile Shield; Pentagon Panel Concurs with Recent Criticism." The story began with these highly misleading words: "A classified report by a Pentagon-appointed panel of experts raises numerous warning flags about the current plan for a missile defense shield...." Only much lower in the article does the reader discover that, "Overall, the new report gives the Pentagon's missile defense developer a 'B plus grade for work done thus far,' and it grants an overall blessing to the plans drawn up for future testing and evaluation...."[98] The headline "Pentagon Panel Gives Missile Defense a B Plus" seems not to have occurred to the editors of the *Washington Post*. The "red flags" and "warnings," meanwhile, turn out to have been merely a matter of timing, not feasibility. "Senior defense officials familiar with the report said it concludes that the complex system of targeting radars, interceptor missiles, and high speed computers eventually should work as designed. But it voices strong skepticism that the system will be operating successfully by 2005."[99]

Democratic congresswoman (now senator) Barbara Boxer of California called SDI "the president's astrological dream...a dream of laser weapons powered by nuclear explosions, particle beam weapons, chemical rockets and space based interceptors parked in 'garages' in orbit."[100] Columnist Mary McGrory sniped, "Let any lunacy come from the White House, Congress will ratify it."[101]

Senator John F. Kerry attempted to strangle SDI in the cradle in 1985, sponsoring a bill that would have frozen spending for the program at $1.4 billion. Kerry's amendment would further have forbidden experiments of any kind that would tend to undermine the "single most important arms-control treaty of our time."[102] He was referring to the ABM treaty—a treaty that the Soviets had violated, and that liberals would continue to worship even after the Soviet Union had ceased to exist. The amendment was defeated 78 to 21.[103]

Writing in the *Washington Post*, columnist Phillip Geyelin scorned SDI as a "telling commentary on [Reagan's] presidential style: Reagan had no proposal worked out when he first floated the idea almost casually in a speech devoted to other, known quantities in his military program. He had only a fatuous, personal vision of a nuclear-free world."[104]

Al Gore, whose 1988 strategy required him to stake out a centrist position, put a smoother gloss on his position, but in substance it differed only slightly from those of the other candidates in the race that year. He said, "I'm sure that President Reagan sincerely believes he is acting with the best of purposes in looking to SDI as a way to permanently end the threat of mutual nuclear annihilation. But he is wrong—wrong because SDI is not feasible, wrong because it would entail exorbitant costs and, most important, wrong because it would make the possibility of a first strike more likely...."[105]

On a different occasion, Gore said:

> There is a group of extremely hard-line conservatives who
> see in SDI a means for destabilizing the arms race by
> deploying defenses to protect our missile silos. This could
> challenge the Soviet Union to an accelerated arms race,
> and, the conservatives hope, pressure the Soviets econom-
> ically to induce a radical change in their system. But their
> strategy is not viable. The Soviets have always found the
> rubles to match our military escalation. We're the ones
> with Gramm-Rudman. To assume that they're the ones
> who would buckle is madness."[106]

Gore's description of the conservative position was a caricature.
Most supporters of SDI hoped to protect the American people, not
just to protect our missiles (though the missiles themselves existed
only to protect people). Gore's crystal ball also turned out to be
extremely cloudy. The Soviets did indeed recognize that they would
go bankrupt attempting to match our SDI program. But the most
interesting thing about Gore's remark was the passing reference to
"our escalation." He shared the view, so widely held on the left, that
it was the United States that "escalated" the arms race, while the
USSR was in the position of merely responding.

The world's only antiballistic missile system at the time surrounded Moscow. And the world's first "killer satellite" system was
deployed by the Soviet Union. Nor did the Soviet Union adopt the
broad reading of the ABM treaty invoked so often by American liberals. In 1972, Marshal Grechko testified to the Supreme Soviet that
"The treaty on limiting ABM systems imposes no limitations on the
performance of research and experimental work aimed at resolving the
problem of defending the country against nuclear missile attack."[107]

The Soviets had more than ten thousand scientists and engineers
and more than a half dozen major research facilities devoted to laser
weapons research. They were studying air defense lasers, antisatel-
lite lasers, and ABM lasers. They had been working on space-based

particle beam weapons since the 1960s, and were exploring the possibility of radio frequency signals that could disrupt or destroy the electronic components of missile warheads.

And while the United States began to dismantle its air defenses in the 1960s, the Soviets bolstered their own. When Reagan first broached the idea of SDI, the Soviets had 12,000 SAM launchers at more than 1,200 sites, 10,000 air defense radars, and 1,200 interceptor aircraft dedicated to strategic defense. An additional 2,800 interceptors could be used in strategic defense missions. This compared with 300 interceptor aircraft on the American side, 118 strategic air defense radars, and zero strategic surface to air missile launchers.[108]

At the universities, anti-SDI sentiment ran high. Decrying SDI as "ill-conceived and dangerous," a group of 6,500 scientists, including fifteen Nobel Prize winners, signed a "pledge of nonparticipation" in SDI research. Dr. Wolfgang Panofsky, a scientist at Stanford University, declined even to dine with President Reagan, explaining that he found the president's call for research on missile defense "somewhat spiritually troubling."[109] Dr. Victor Weisskopf of MIT did attend the dinner but emerged denouncing the idea of SDI as "extremely dangerous and destabilizing."[110] Robert McNamara dismissed the concept as "pie in the sky."[111] Appearing on the *Donahue* program, Ted Koppel, who had described those who favored SDI as "hardline,"[112] was asked to elaborate. Koppel cautioned that he could not be uninhibited in his comments: "I have to maintain some kind of objectivity." But he did allow that "I think what is being proposed for expenditure on Star Wars [sic], for example, is absolute nonsense. Anything like an SDI program is going to put us in a position where, naturally, the Russians are going to feel threatened." To complete the usual "It's-Dangerous-and-it-Won't-Work" mantra, Koppel finished by adding, "There is no way that it is going to work within the next twenty years and it is going to cost not billions, not tens of billions, not hundreds of billions, but trillions of dollars."[113]

On ABC's *World News Tonight*, anchor Peter Jennings called missile defense a system "that has never been proven to work and may never work."[114] The *New York Times* took a hiatus from its usual tone of heated indignation at the idea ("a pipe dream, a projection of fantasy into policy")[115] to adopt a bit of lofty condescension: "For ages, nations have dreamed of building invulnerable shields to protect themselves from hostile forces. There was the Great Wall and the Maginot Line and, in America the safeguard antimissile system and 'Star Wars.' All fell short."[116]

Not exactly. Defenses have been successful for as long as humans have fought wars. That's why they were used. People are not complete fools (though you might not always be able to recognize this from the editorials of the *New York Times*). The shield helped to stave off the club, and chainmail helped to blunt the effect of arrows, as did slant windows in battlements. Boiling oil proved an effective defense of castle walls against would-be scalers using grappling hooks, as were moats. Radar proved an invaluable defensive asset to Great Britain in the Battle of Britain. And shoulder fired Stinger antiaircraft missiles may have tipped the balance against the Soviet Union's HIND helicopters in Afghanistan. In time, every offense provokes a defense and vice versa. But whereas the *New York Times* and other liberals cite this history to prove the folly of investing in missile defense, that is not at all the proper lesson. Castles and drawbridges kept Europeans comparatively safe for generations. Nothing lasts forever. But in human terms, a couple of centuries is a long time.

Some liberals worried, along with the Soviets, not that SDI would fail, but that it would succeed. The Soviets feared that with an effective defense against ballistic missiles, the U.S. would have complete freedom of action around the world and the Soviet nuclear arsenal would become "impotent and obsolete." That the Soviets feared such an outcome requires no explanation. And at Reykjavik, Gorbachev would offer Reagan sweeping arms-control concessions in return for a treaty keeping SDI "in the laboratory."

What is not self-explanatory is why liberals feared SDI's success. Certainly an effective missile defense in the hands of the Soviets would be a fearful thing (and they were working hard to build one). But in the hands of the United States, how could it be anything but a force for good? We, unlike the Soviets, had no hopes for world domination, no plans for expansion, no repressive regime to impose on others, and no lust for conquest. President Reagan repeatedly offered to share the technology with the Soviets (though it is true that Gorbachev never believed this,[117] and some Americans thought it quite unwise).

But the point is: a weapon in and of itself is neither dangerous nor harmless. Everything depends upon the intentions of the weapon's owner. This is the answer to those who argue that the United States is "hypocritical" to oppose the proliferation of nuclear weapons to states like Iraq and Iran, while turning a blind eye to Israel's possession. Hypocrisy has nothing to do with it. We know that Israel is not an aggressor and would use nuclear weapons, if ever, only in desperate self-defense against enemies sworn to her destruction. Iraq and Iran are another story.

Democratic presidential candidate Michael Dukakis described SDI as "a fantasy—a technological illusion which most scientists say cannot be achieved in the foreseeable future. The defenses they envision won't make the United States more secure—they will simply fuel the arms race, as each new system produces a counter system, with no increased security. Deploying defenses could make nuclear war more, not less, likely...."[118]

Liberals were in love with this particular critique of SDI. Thomas Friedman, a *New York Times* columnist, used a sartorial metaphor to debunk the second Bush administration's plans for missile defense. "It's good to have layers of defense, just as it's good to have belts and suspenders. But if you already have suspenders, it would be crazy to pay $100 billion for a belt of uncertain reliability—especially if that belt makes it more likely your pants will fall down."[119]

There it is, the perennial suggestion that building SDI would increase the likelihood that the Soviet Union would launch a first strike. Friedman never explained how adding even a faulty belt could make a pair of pants held up by working suspenders *more* likely to fall down. His metaphor only holds together if we assume that the belt was made of lead.

The arguments of SDI opponents were self-contradicting. They said that SDI was a fantasy that would never work and at the same time that it was dangerous and destabilizing. How could it be dangerous if it would never work? If it was quite the joke critics portrayed it, why weren't the Soviets laughing?

Perhaps because they understood that SDI could work, and further that they could never hope to compete with the U.S. in the fields that SDI would require—computer science and engineering. Former Soviet ambassador to the U.S., Anatoly Dobrynin, confirms that the Soviets regarded SDI as a "real threat" because "our leadership was convinced that the great technical potential of the U.S. had scored again."[120] Fear of U.S. strength, particularly of American technical superiority, drove the Soviets to desperate measures. Along with a fierce propaganda campaign decrying SDI as the worst threat to world peace in history, they offered remarkable arms concessions at the hastily arranged Reykjavik summit in Iceland. Gorbachev promised drastically to reduce Soviet conventional forces in Europe, to eliminate intermediate range missiles in Europe, and to reduce their nuclear arsenal by half. The only catch was that the United States would have to agree to limit SDI research to the laboratory. Reagan was furious. In his memoir, he described the moment:

> [Gorbachev] wouldn't budge from his position. He just sat there smiling and then he said he still didn't believe me when I said the United States would make the SDI available to other countries.
>
> I was getting angrier and angrier.
>
> I realized he had brought me to Iceland with one purpose: to kill the Strategic Defense Initiative.[121]

On July 14, 2001, the military performed its fourth test of SDI technology. A dummy warhead was blasted out of the sky by a missile launched from 5,000 miles away. The dummy was destroyed in mid-flight, when it and the interceptor were traveling at a combined speed of 16,200 miles per hour. A bullet had hit a bullet—just as Reagan had supposed would be possible. Just as Gorbachev had feared. "Missile Test Success Raises Hope, Anxiety" read the *Boston Globe* headline a couple of days later, "Allies Concerned Over System, But U.S. Critics Quieted."[122]

If only it were true. In fact, "U.S. critics" never admitted that they were wrong. They seldom do.

CHAPTER SIX

# Each New Communist Is Different

*Then the idiot who praises with enthusiastic tone,*
*All centuries but this and every country but his own.*
— GILBERT AND SULLIVAN, *The Mikado*

THOUGH THE SOVIET UNION PROVED a disappointment to the America-despising Left, a long series of pretenders to the crown of egalitarian utopia presented themselves during the decades following World War II. If the Soviet Union had sunk into show trials and mass killings, China would herald the new dawn. When Maoism marched inexorably into barbarism (on a scale that dwarfed even the Soviet and Nazi genocides), doctrinaire Marxists looked to the next and latest incarnation of the Communist idea. Even weird, remote North Korea won some adherents among hard leftists. In his memoir, *Commies,* Ronald Radosh, a former leftist, recalls meeting Robert Scheer, who would later become a reporter for the *Los Angeles Times* and then a syndicated columnist. In 1970, Radosh recalls, Scheer regaled an audience on Pacifica radio about the glories of Kim Il-Sung's regime:

> For over two hours, Scheer talked and talked about the paradise he had seen during a recent visit to North Korea, about the greatness of Kim Il-Sung, about the correct nature of his so-called *juche* ideology—evidently a word embodying Kim's redefinition of Marxism-Leninism in building communism against all obstacles and with the entire world in opposition. Others in the Movement had, of course, found heaven on earth in places like Cuba and even China.... Scheer had one-upped them all by discovering a paradise in Pyongyang.[1]

But Scheer was unusual. China and North Korea were on the other side of the world. Only the most committed Communists and fellow travelers would make the trek. (It is worth pausing to notice that Scheer, so recently a warm admirer of North Korea, is almost universally regarded as an American liberal.) The more popular destinations for what Paul Hollander called "political pilgrims" were Cuba, Grenada, and most popular, Nicaragua.

The drama unfolded almost identically with each new socialist paradise. First, liberals would denounce the shortcomings and corruption of the noncommunist government under siege by revolutionaries. Next, they would deny that the rebels were Communists, preferring to believe (or certainly to claim) that their leadership was "mixed" or a coalition of opposition groups. Communists did often make common cause with other opponents of the regime but once in power jettisoned or killed their erstwhile allies. Was it beyond the capacity of starry-eyed liberals to notice this? When the Communist regime, safely in power, began to militarize, close down independent newspapers, and collaborate with the Soviets, liberals would lament that U.S. hostility drove the "agrarian reformers" into the Soviets' arms. One example among thousands is a speech by Democrat senator Paul Tsongas of Massachusetts on the floor of the Senate in 1980: "Castro is in power today because we isolated him and gave him no choice but to turn to the Soviets."[2] Castro himself peddles this same line ceaselessly. That Castro was a zealous Communist long before the United States turned on him has been long since established. Today, however, Castro grinds out the fiction that nothing but the U.S. embargo keeps his island from economic bliss.

## COMMUNISM WITH SEX APPEAL

The Cuban revolution began as a victory for anti-Batista forces including the "26 of July Movement." For tactical reasons, Castro sent mixed signals about his doctrinaire Marxism, correctly calculating that the sympathy of American liberals could work to his advantage. When Castro appeared on *Meet the Press* in April 1959, Lawrence Spivak posed this question: "I want to know where your heart lies in the struggle between communism and democracy. Whose side, where is your heart and where are your feelings?" Castro lied. "Democracy is my ideal, really....I am not communist. I am not agreed with communism....There is no doubt for me between democracy and communism."[3]

Of course, he needn't have lied to obtain the support of some American academics and intellectuals. The more radically left Castro was, the better they liked it. C. Wright Mills of Columbia University extended unqualified support to Castro declaring, "I am for the Cuban revolution. I do not worry about it. I worry for it and with it."[4] The Venceremos Brigade and the Fair Play for Cuba Committee swallowed—and spouted—the party line. Because Castro's bid for liberal support came before America's Vietnam vortex, he could not count on the undiluted sympathy of anyone not already aligned with the far Left. Still, Castro made a much-discussed tour of the United States in 1958, visiting Washington, New York, and Boston, where he delivered a lecture at Harvard. His bravado and swagger made many an intellectual weak in the knees. Novelist Norman Mailer was typical. Still swooning after Castro's visit to New York, Mailer wrote:

> So Fidel Castro, I announce to the City of New York that
> you gave all of us who are alone in this country...some
> sense that there were heroes in the world. One felt life in
> one's overargued blood as one picked up in our newspa-
> pers the details of your voyage....It was as if the ghost of
> Cortez had appeared in our century riding Zapata's white
> horse. You were the first and greatest hero to appear in the
> world since the Second War....you are the answer to the
> argument of Commissars and Statesmen that revolutions
> cannot last, that they turn corrupt or eat their own.[5]

Senator George McGovern was naturally disposed to believe the
best of the Cuban dictator. "The reality of the man matches the
image," he recorded in his memoir. "He was, as always, dressed in
freshly pressed military fatigues. Youthful in appearance at the age of
48, hair and beard still black, cigar constantly in hand, he was poised,
confident and questioning....He responds knowledgeably on almost
any subject from agricultural methods to Marxist dialectics to American politics."[6] McGovern left the island convinced that "Castro has
the support and outright affection of his people."[7]

Jonathan Kozol, the author of a number of books on American
education, was a full-throated enthusiast of Cuba's schools:

> There are these words in the Bible: "Where there is no
> vision, the people will perish." In Cuban schools...the
> vision is strong, the dream is vivid and the goal is clear.
> There is a sense, within the Cuban schools, that one is
> working for a purpose and that that purpose is a great deal
> more profound and more important than the selfish plea-
> sure of individual reward. The goal is to become an active
> member in a common campaign to win an ethical objective.[8]

And so on. Cuba was more humane, more authentic, more
vibrant, less inhibited, and in every important respect superior to the
United States. So claimed Angela Davis, Jean-Paul Sartre, Todd Gitlin,

Susan Sontag, David Dellinger, Saul Landau, Julius Lester, and many more. Barry Reckord acknowledged that not everything in Cuba was ideal but the flavor of the enterprise was so life affirming that small failings could be forgiven:

> The individual who is unjust works in a society that shares its food fairly evenly, doesn't discriminate against blacks, gives priority to hospitals and schools over Cadillacs and mansions, wipes out unemployment with the money it borrows. The society is moral even if the individual is not. Men are more just, more humane in a collective sense.[9]

Che Guevara, his legend enhanced by an early death in 1967, added movie star good looks to the Communist movement—a quality not in large supply among revolutionaries before then. Che's photo adorned thousands of college dorms during the heyday of the New Left in the 1960s and early 1970s. The man lionized by so many baby boomer liberals was born into a wealthy Buenos Aires family. Che met Castro in Mexico in 1955 and returned to Cuba with him to advance the revolution. "I am one of those people," he said, "who believes that the solution to the world's problems is to be found behind the Iron Curtain." Though members of the New Left would often claim that they did not share the Old Left's affection for the Soviet Union, their admiration for Communism was really just once removed. They did not admire Lenin, but they worshipped Che, who named his first son Vladimir in Lenin's honor.[10]

I. F. Stone was an American journalist much admired by liberals of every stripe. A die-hard Communist sympathizer, he never stepped all the way over the line to Stalinism—at least not after 1956 and Khrushchev's famous "secret speech." But he came quite close. After meeting Che Guevara, he swooned like a teenager:

> He was the first man I had ever met whom I thought not just handsome but beautiful. With his curly, reddish beard,

> he looked like a cross between a faun and a Sunday school print of Jesus....
>
> In Che, one felt a desire to heal and pity for suffering.... It was out of love, like the perfect knight of medieval romance, that he had set out to combat with the powers of the world.... In a sense he was, like some early saint, taking refuge in the desert. Only there could the purity of the faith be safeguarded.[11]

Che Guevara was a cruel fanatic. After Castro made him a commander of a resistance unit, he gained a reputation for harshness. A young boy in his unit was caught stealing a bit of food. Without a trial, the boy was shot on Che's orders. After the revolution's success, Che became state prosecutor. In perfect repetition of the Soviet model, he carried out death sentences on many men, including former comrades who refused to shed their belief in democracy. Guevara also holds the distinction of establishing Cuba's first forced labor camp. In his will, he praised the "extremely useful hatred that turns men into effective, violent, merciless, and cold killing machines."[12]

Such was the power of the Communist idea that sympathizers truly did call black white and good evil. How they could also claim the moral high ground, however, remains a mystery of the twentieth century.

Ron Radosh recounts traveling to Cuba with other leftist sympathizers in the 1960s:

> ...While the Cubans were trying to squeeze into overcrowded buses in the August heat to get to jobs where they had to work an average twelve-hour day, my comrades and I enjoyed a lobster and shrimp luncheon in the best hotel in Cuba, the Havana Libre, formerly the Havana Hilton, built the year before Castro's victory. There I drank wine... in the sacred presence of Che Guevara's widow and other members of his family.[13]

In the course of the visit, despite their Communist tour guide's manly attempts to obscure reality, Radosh at least was able to perceive that the Cuban people were suffering. Most of the other members of the delegation, however, chose to check their consciences at the door. After visiting a Cuban mental hospital, members of the group were a little dismayed to learn that lobotomy was a widely practiced "therapy" for mental patients (long after the practice had been abandoned in the U.S.). In fact, most of those committed to mental hospitals in Cuba were lobotomized. One member of the group protested, saying, "This stinks. Lobotomy is a horror. We must do something to stop this. It's exactly what we're working against at home." But another pilgrim, Suzanne Ross, was unperturbed: "We have to understand that there are differences between capitalist lobotomies and socialist lobotomies."[14] A fitting motto for fellow travelers at all times and places.

A similar refrain was sung by church leaders. Bishop James Armstrong and the Reverend Russell Dilley of the United Methodist Church visited Cuba in 1977. Apparently unable to deny the existence of political prisoners, they wrote, "There is a significant difference between situations where people are imprisoned for opposing regimes designed to perpetuate inequalities (as in Chile and Brazil, for example) and situations where people are imprisoned for opposing regimes designed to remove inequities (as in Cuba)."[15]

Castro's regime, one of the last Communist holdouts, is a communist tyranny. Religion is suppressed and the pious are persecuted, often cruelly. There is no independent newspaper, court, church, business, union, or book club. Homosexuals are publicly humiliated and often imprisoned. Armando Valladares, who has been rightly called the "Cuban Solzhenitsyn," spent two decades in Castro's Gulag. He was arrested in 1961 for reasons he could hardly guess at, but during his first interrogation he learned that he had been branded a counter-revolutionary because he had expressed misgivings about communism. For this he was subjected to twenty nightmarish years in

Castro's prisons. Early in his tour, he met Clodomiro Mirada, one of Castro's former comrades who had fought Batista in the mountains of Pinar del Rio. But Mirada was fighting for freedom, and when he saw that Castro was perverting the revolution toward communism, he took to the hills again as a revolutionary. Valladares describes the end of his life:

> Castro ordered him hunted down, and thousands of militia were sent out to find him. He was wounded in a skirmish. When they captured him later, his legs had been completely destroyed by bullets, and there were other shells lodged in one arm and one side of his chest. He was carried into his trial on a stretcher. When they sentenced him to death, he was taken out of the military hospital and locked up in one of those horrific cells without a bed. Clodomiro was unable to stand up, so he had to drag himself along the filthy floor. His unattended wounds became infected; then they filled with maggots. That is how Pedro Luis and Manuel Villanueva found him. They were the last prisoners to speak to him.
>
> It was also on a stretcher that they took Clodomiro down into the [dry] moat to the firing squad. The stairway, which descends into the moat, hangs from the wall on one side. On the other side, there is just space—not even a handrail. The two-hundred-year-old stone steps, worn down by generations of slaves and prisoners, can be seen even from the end of the galleries. The file of guards which carried Clodomiro descended unsteadily. Almost at the bottom of the stairs, one of the guards stumbled. He let go of the stretcher as he groped to steady himself, and Clodomiro fell onto his wounded legs and tumbled down the last steps. One of the guards told us that they tried to tie him to the post, but he simply couldn't stay erect. They

had to shoot him as he lay on the ground. When they shot him, he too cried "Down with communism!"

Clodomiro was perhaps the only man ever executed who was being devoured by worms even before he died.[16]

Soon after Castro took power in Havana, labor leaders and newspaper editors (including those who had vigorously opposed Batista) were exiled or arrested. Religious schools and universities were closed.[17]

Poets, artists, and intellectuals who did not acknowledge that "the revolution is all; everything else is nothing" were persecuted. Many left the country, others were imprisoned. Torture, malnutrition, and execrable living conditions characterized Castro's jails. Prisoners were tortured with sleep deprivation (a Soviet import), beatings, denial of medical attention, being forced to climb a flight of stairs with weighted shoes and then being thrown down the stairs to climb again, and exposure to extreme heat. Almost as if taking a leaf from George Orwell's *1984*, the Communists would sometimes play upon their prisoners' phobias. A woman with a fear of insects was placed in a cell infested with cockroaches.[18]

Cuba has been an island prison since 1959, but living conditions plunged even further after the demise of the Soviet Union. Without the USSR's infusion of $6 billion per year, ordinary Cubans have endured a downward spiral. As Richard Grenier reported after a visit during the early 1990s:

> Cubans have no soap, no detergent, no cooking oil... hardly any meat, or chicken or fish... [and] only one tiny loaf of bread a day. Absolutely everything is rationed, and just because your ration book says you're entitled to something doesn't mean much. Stores are empty. Electrical power has been cut by half. Cuba has blackouts, no air conditioning, no night baseball, no gasoline for busses, almost no cars, and few bicycles. Oxen till the fields. Even

new windmills have broken down for lack of spare parts. It's been many years since Cuban women have had sanitary napkins. They use rags.[19]

Still, when the NBC *Today* show with hosts Bryant Gumbel and Katie Couric visited Cuba in 1992, they mouthed some of the chirpy nonsense that has always characterized liberal views of Cuba. Ms. Couric made approving noises about the "terrific health-care system" and praised the standard of living as "very high for a Third World country."[20] In point of fact, Cuba's standard of living fell relative to other Caribbean nations after Castro's coup. A nation that had been among the wealthiest in Latin America took its place among the poorest. The people of Mexico, Jamaica, the Dominican Republic, and Colombia enjoyed higher standards of living than Cubans.[21]

Writing in the *Washington Post*, assistant foreign news editor Don Podesta opined that:

> If nothing else, the Cuban revolution has eliminated abject need. The cost may be generalized poverty and zero political pluralism, but, even with shortages, there is no starvation here. Education and medical care are assured for all. And unlike in most of Latin America, you don't see naked or even shoeless children in the streets. When Castro speaks of the need to defend the gains of revolution, he means a level of social welfare rare in the underdeveloped world.[22]

*Time* magazine's Cathy Booth chimed in with, "Young Cubans increasingly see themselves as the last idealists in a world that cares only about money.... Ninety miles away in Miami, Cuban émigrés wish for Fidel's imminent collapse, but the island's university students who volunteer to take a two-week 'vacation' in the fields don't see trouble brewing in Paradise."[23]

Another example: Kathleen Sullivan, the host of CBS *This Morning*, told millions of Americans in 1988 that "Now half of the Cuban population is under the age of twenty-five, mostly Spanish speaking, and all have benefited from Castro's Cuba where their health and their education are priorities."[24] On another occasion Sullivan mentioned Cuba's "model health care system."[25] Peter Jennings quite often found reasons to praise Cuba. In 1989, as the rest of the communist world was reaching eager hands toward freedom, Jennings continued to laud the "accomplishments" of Castro's Cuba. "Medical care was once for the privileged few," Jennings told ABC viewers, "today it is available to every Cuban and it is free. Some of Cuba's health care is world class. In heart disease, for example, in brain surgery. Health and education are the revolution's great success stories."[26]

This canard has been repeated more often than the notion that Mussolini made the trains run on time. But when Castro goes, the truth will come out. The vaunted Cuban health care system will be revealed for what it is: a two-tiered system in which the elites, party members, Castro's inner circle, and cash-paying "health tourists" from Europe and elsewhere receive quality care whereas the ordinary Cuban settles for long waiting lists, poor supplies, shortages of necessary medicines, and crumbling clinics.[27]

The credulity of American reporters in Cuba—even to the present day—is staggering. They accept, at face value, the government's claims about health and literacy. But as we have seen in the case of the Soviet Union, Communist governments do not just inflate their statistics, they invent them.

In 2000, ABC correspondent Cynthia McFadden offered this account of a Havana second grade classroom:

> Part of what the children talked about was their fear of the United States and how they felt that they didn't want to come to the United States because it was a place where they kidnap children, a direct reference of course, to Elian Gonzalez. The children also said that the United States was just a place where there was money and money wasn't what was important. I should mention, Peter, that . . . as you talk

about the global community, Cuba is a place because of the small number of computers here—in the classrooms we visited yesterday there were certainly no computers and almost no paper that we could see—this is a place where the children's role models and their idols are not the baseball players or Madonna or pop stars. Their role models are engineers and teachers and librarians—which is who all the children we spoke to yesterday said they wanted to be.[28]

If some hint of scorn might be detected about the indoctrination that these kids had obviously endured on the subject of Yanqui kidnapping, it was muted. (Elian Gonzalez, of course, was not kidnapped, but fled to the United States with his mother.) The American audience was expected to accept that, in contrast to the crass American adulation of pop stars and athletes, Cuban children spontaneously express their admiration for engineers, teachers, and librarians. This is a hardy perennial. In the 1930s, it was said that Soviet workers spontaneously hailed "Stakhanovite" workers. (Alexei Stakhanov was a Russian coal miner who supposedly performed great feats of productivity. Stalin used him as a model for all Soviet workers. As it turned out, the Stakhanov story was a fraud. No surprise there.) As for the shortage of computers and even paper—well, "money isn't what's important," though if Ms. McFadden were reporting on an American school (or a school in a nation allied with the U.S.) found to be so ill equipped, it would surely have been reported as a scandal, not as evidence of a more enlightened, nonmaterialistic ethic.

Jennings added his own benediction: "From the Cuban point of view, as everybody knows I guess, education and participation in the Third World are very much what Cuba has stood for, at least in the developing world."[29] On April 3, 1989, Jennings summed up Castro's thirty years of dictatorship by declaring, "Castro has delivered the most to those who had the least. And for much of the Third World, Cuba is actually a model of development.... Education was once available to the rich and the well-connected; it is now free to all. On January 1, 1959, when the Cuban dictator [Fulgencio] Batista left the country for good, only a third of the population could read and write. Today the literacy rate is 97 percent."[30]

Jennings comfortably identified Batista as a dictator, but somehow that word has never crossed his lips regarding the man who has held power for forty-four years without an election. Jennings, like so many others who ought to know better, continues to accept uncritically the official government statistics provided by the Castro regime.

As Nicholas Eberstadt, author of *The Poverty of Communism*, points out, Cuba's literacy rate had been high, 76 percent, when Castro seized power in 1959. Even if one assumes a literacy rate of 90 percent today (accepting, arguendo, the Communist government's numbers), that is still less progress than was achieved by Cuba's noncommunist neighbors in Latin America during the same period.[31] And there is good reason to believe that Cuba's infant mortality statistics have been falsified.[32]

In 1981, a United Methodist Church document described Cuba as "a vision for the future."[33] Liberals are comfortable believing that Communist countries have delivered on their promises of better material lives ("a model health care system!" "universal literacy!"), but can easily switch gears when confronted by the poverty and deprivation Communism actually delivers. Michelle Singletary, writing in the *Washington Post*, did not deny the poverty she saw in Cuba, but rationalized it this way:

> In Cuba there are no shelves full of Barbie dolls. There is no Disney World. Instead of aerodynamic skateboards or sparkling Rollerblades, many Cuban children are forced to fashion their own toys. I watched as three young boys darted around traffic on makeshift scooters made out of old crates. Just down the street, other boys were playing drums on empty cardboard boxes.

...So many of us in America live what Cubans would consider very prosperous lives. Yet we worry that we don't have enough while our homes are filled with gadgets and things paid for with money we don't have. We shower our children with so much stuff that there is always a perpetual layer of toys in their pricey toy bins that they never play with again.[34]

How many families in Communist and other totalitarian countries would cherish the chance to raise their children in an environment of plenty—where the burden falls upon parents, not state-imposed poverty—to prevent children from being spoiled. But to a left-leaning American journalist, economic deprivation in a Communist country is not a sign of failure, but instead of moral and psychological superiority.

In 1997, CNN became the first Western news organization to open a bureau in Havana. Those who hoped for an accurate glimpse of what life is like for ordinary Cubans were disappointed. As the Media Research Center reported after studying five years' worth of dispatches from Havana, CNN reported on political prisoners and dissidents only seven times. As Brent Bozell, MRC's president, told the National Press Club, "that's fewer than half as many stories as CNN produced in just the first three months of 2002 about phony claims of human rights abuses [committed] by the United States against those held at its base in Guantanamo Bay, Cuba."[35] CNN also gave spokesmen for the Cuban regime six times as much airtime as noncommunist spokesmen like church leaders or dissidents. Similarly, when CNN presented the views of ordinary Cubans, they broadcast six interviews with those who supported the regime for every one with those who did not.[36]

Dissidents inside Cuba have attempted to use CNN's cameras to publicize their plight and their cause—often with disappointing results. Though members of the press expend great efforts on behalf of their fellow journalists who face persecution in other parts of the world, Castro's repression of speech is permitted to go on in darkness. Several hundred protesters gathered in November 2000 at the home of Jose Orlando Gonzalez Bridon, a leader of the illegal Cuban Democratic Workers' Confederation, a would-be trade union. A CNN reporter and camera crew were present, yet CNN never aired the footage. CNN denied that their reporter had ever promised to televise the protest, but the protesters who had risked a great deal to be there and who took that risk only for the chance to see their cause elevated by CNN, were bitter. Later, many of those who attended the protest were beaten and jailed.[37]

The other networks were similarly astigmatic in their views of Cuba. During Gorbachev's visit to Cuba in 1989, ABC News aired thirty-seven stories—and only one dealt with human rights violations. CBS reported on the arrest of a score of human rights activists, but ABC never mentioned them, preferring to fill its airtime with praise of Cuba's health and education system.[38]

One would never guess from American news reporting that Cuba was the kind of country that hundreds of thousands of people would rush to flee at the merest flicker of opportunity. Over a million Cubans fled to the United States during the 1960s, "voting with their oars." That represented one in nine inhabitants of the island.[39] Only the most intolerable conditions will drive people to leave their homes and risk their lives to escape. In one Communist country after another, millions have fled. And yet some who think of themselves as enlightened and humanitarian on the American Left have found ways to ignore or excuse this damning fact.

In 1980 a rumor swept Havana that the Peruvian embassy was offering visas. The embassy grounds were swamped as desperate Cubans scaled the embassy compound walls hoping to escape. Eventually, Castro permitted 125,000 to depart from the port of Mariel—though he put his finger in America's eye by sending along criminals and the mentally ill as well. In the year 1994, forty-one Cubans died attempting to make it to Florida in small boats.[40]

Still, the Congressional Black Caucus adores Fidel Castro. New York Democrat Charles Rangel greeted Castro with a bear hug in Harlem during the latter's visit to New York in 1995. Audiences there greeted the dictator with chants of "Fidel, Fidel, Fidel." Jesse Jackson traveled to Cuba and praised the Communist leader before a cheering crowd at National University in Havana—"Long live Cuba! Long live the United States! Long live President Castro! Long live Martin Luther King! Long live Che Guevara! Long live Patrice Lumumba! Long live our cry of freedom! Our time has come."[41] Jackson and Castro together then received the adulation of the crowd.

Castro has always played the race card in relations with the United States and has courted black Americans skillfully. He has also made his island a refuge for criminals who style themselves "political prisoners" in the U.S. One such was Joanne Chesimard, who in 1973 murdered a police officer in New Jersey. She was sentenced to life in prison, but escaped in 1979, changed her name to Assata Shakur, and made her way to Cuba. There she dines out on her "persecution" by the American criminal justice system, regaling Latin American and European visitors with her experiences. At the same time, people whose only crime was to attend Mass or try to publish a free newspaper languish unlamented in Castro's jails.

As Jay Nordlinger reported in *National Review*, Shakur was the subject of a misunderstanding between Fidel Castro and California Democratic congresswoman Maxine Waters. Waters had voted for a measure calling for the extradition of "Joanne Chesimard." Only later did Waters discover that Chesimard was Assata Shakur. Abashed, Waters wrote a truckling letter to Castro personally explaining that she had no idea these women were one and the same and would never have voted as she did if she had known. She went on to decry the "Republican leadership" in the House of Representatives for "deceptive intent" in using the old name, and added that the 1960s and 1970s had been "a sad and shameful chapter of our history" when "vicious and reprehensible acts were taken against" people like Shakur, forcing them to "flee political persecution."[42] It's difficult to say which part of this letter is the most disgusting: the "apology" by a sitting member of the United States Congress to a brutal dictator; the suggestion that it was somehow "vicious and reprehensible" for the U.S. to convict and imprison a murderer; or the notion that a country like Cuba was a refuge from "political persecution" when, in fact, Cuba is one of the world's leading practitioners of political persecution?

Cuba also remains a pet cause of the National Council of Churches. In 1975 the NCC published a pamphlet entitled "Cuba: People-Questions." As Tucker Carlson observed in the *Weekly Standard*, it was written in "perfect irony-free Albanian farm report prose." Purporting to offer a short history of U.S.-Cuban relations, the pamphlet explained, "All through the 1960s, the U.S. did its best to make Cuba buckle under." With the use of an economic blockade, "cold war tactics," and a "CIA-led army to act against the revolutionary government," the U.S. bullied the island nation. Fortunately, the Cuban people "overwhelmed the invaders" at the Bay of Pigs and the revolution marched on. "Later on the leaders are to call that socialism. The poor people call it great."[43] The pamphlet also praised the example Cuba was setting for the rest of the world, noting that "guerrilla and other grass roots movements" are "drawing courage from Cuba."[44] And indeed they were.

## GRENADA: RORSCHACH TEST

In the late 1970s, when the United States was still febrile with post-Vietnam syndrome, revolutionaries in Central America and the Caribbean pushed their advantage. In El Salvador, the FMLN (Farabundo Marti National Liberation Front), a Marxist guerrilla army stepped up its war against the noncommunist central government. In Nicaragua, assisted by Fidel Castro, the Sandinista Liberation Front was able to seize power when the dictator, a smarmy despot named Anastasio Samoza, was ousted by a coalition of opposition

188 | Useful Idiots

groups. And in tiny Grenada, a small band of hard-core Marxists took control of the island and put it at the service of Cuba and the USSR.

Marxist Maurice Bishop and the New Jewel Movement unseated Eric Gairy in a 1979 coup d'etat. Grenada immediately smiled in the direction of the USSR, Cuba, Libya, and North Korea. Plans were put into place to build a ten-thousand-man army—about 10 percent of the island's population. In addition to representatives of various Eastern European Communist regimes, Grenada was host to more than a thousand Cuban troops, and had signed military agreements with the USSR, Cuba, and North Korea. A senior general from the USSR was on hand.

In October 1983, competitors within the New Jewel Movement murdered Bishop and the island was plunged into chaos. A twenty-four-hour shoot-to-kill curfew was imposed. The presence of about one thousand Americans in Grenada, including seven hundred medical students, alarmed the American government. Responding to an urgent request from six Caribbean nations—Jamaica, Barbados, St. Vincent, St. Lucia, Dominica, and Antigua—the Reagan administration launched Operation Urgent Fury, a military invasion to free the endangered Americans and restore order and democracy to Grenada.

Most of the world condemned the U.S. action. The USSR predictably accused the United States of aggression and lawlessness—though the Soviets displayed pictures of Grenada, Spain, not the Caribbean island, on their nightly news program.[45]

The operation was a success. The freed American students kissed the ground when they arrived back in the United States (to the consternation of Reagan's liberal critics), and Grenadians were thankful to be free of their oppressive government. One of the documents seized after the invasion was a memo from the New Jewel "politburo" noting that in four short years they had dragged the once comfortable nation into bankruptcy. The economy was so weak that the nation could no longer meet the requirements for aid from the International Monetary Fund. A revealing notation suggested "using the Surinam and Cuban experience in keeping two sets of records in the banks for this purpose.... The comrades from Nicaragua and Cuba must visit Grenada to train comrades in the readjustment of the books."[46] The Cuban presence on Grenada was keenly felt. Castro had called his detachment "construction workers." If so, American troops later reported back, they were using "automatic repeating shovels."

The American invasion, a small and relatively tidy affair as military actions go, freed the American students, drove out the Communists and their Cuban patrons, and provided for free elections within weeks of the American departure. American troops found enormous stashes of heavy weaponry as well as plans to transform the island into a military base for the Communist powers. A success all around? Not quite. The invasion, the first use of U.S. military power in Latin America since 1965, proved a Rorschach test for American liberals. While large majorities of Americans supported the action, liberal opinion was nearly apoplectic. Liberals demonstrated, by their outrage, that their descent into McGovernism endured.

The *New York Times* could discern no moral difference between this U.S. action, intended to free endangered Americans and restore order and freedom, and the Soviet invasion of Afghanistan, which was intended to subdue an entire nation and make it the servant of Soviet-installed masters. "Simply put," said America's preeminent newspaper, "the cost is loss of the moral high ground: a reverberating demonstration to the world that America has no more respect for laws and borders, for the codes of civilization, than the Soviet Union."[47]

Democrat senator Carl Levin of Michigan said:

> There is no legitimate reason for the United States to seek to overthrow other governments we don't like. The Soviet Union has no right to impose its will upon the people of Poland. How then, at the same time, can we insist that we have a right to impose our way of life upon another people in this hemisphere?[48]

Representative Ted Weiss, a New York Democrat, saw Levin's Afghanistan and raised him: "In ordering the invasion of Grenada, Ronald Reagan has adopted the tactic of the Japanese attack on Pearl

Harbor as the new American standard of behavior."[49] California Democrat (now a senator, but then a representative) Barbara Boxer taunted, "This is Wednesday, and we must be in Grenada or Nicaragua or Lebanon or God knows where tomorrow. If we follow the reasoning put forth by many of my colleagues on the [Republican] side of the aisle, we may well be cheering on American forces in dozens of countries all over the world."[50]

Democratic congressman (later senator) Robert Torricelli of New Jersey warned:

> We have entered a rising spiral of conflict the world over. The hour is late. One morning we will wake to hear that a young American and a young Soviet have faced each other in one of these lands, a shot will have been fired, and the world will bleed. Mr. Reagan, today, if you listen ... meet Mr. Andropov to discuss Grenada, Lebanon, or whatever you seek. But Mr. Reagan, try by all means to save the peace before it is too late.[51]

*Washington Post* columnist Richard Cohen lamented that for "thirty-eight years, we have fought repeatedly and, it would seem, to no avail. The dreaded 'Soviet menace' advances. We lost in Vietnam, tied in Korea, and the world has gotten no safer.... By using force to exercise control, you stand a good chance of losing control."[52] Note that Cohen placed the words *Soviet menace* in quotation marks.

Democrat congressman Pete Stark of California had the same general idea but expressed it with his characteristic judiciousness and sobriety:

> It is essential now that the president has shown his true colors that the Congress take control of the situation ... and bring this insane Reagan foreign policy back into line.... [T]he president has been itching to try something like this ever since he was elected. He loves to throw American

weight around, and where better than in a small island with no armed forces within our own hemisphere.... We are rapidly headed down the road not to one Vietnam, but possibly two or three, and maybe even World War III. Let us stop the president before it is too late.[53]

Jesse Jackson proclaimed: "We ought to be outraged.... I call for an immediate cessation of military action against Grenada and reparations to the Grenadian people for damages caused by this invasion."[54] And Democratic senator Pat Leahy of Vermont chose ridicule: "We'd like to have another country to invade, but they can't find one small enough."[55] (Leahy was forced to resign from the Senate Intelligence Committee in 1986 after it was revealed that he had compromised confidential documents by leaking them to the press.)[56]

Walter Mondale, then the front-runner for the Democratic presidential nomination, sounded the liberal mantra as well:

> They [the Reagan administration] moved ahead of the facts, and in violation of the most fundamental of all principles: nonintervention of one state in the affairs of another. And I'm deeply concerned about it.... [Grenada] undermines our ability to effectively criticize what the Soviets have done in their brutal intervention in Afghanistan, in Poland, and elsewhere.[57]

Was "nonintervention of one state in the affairs of another" the "most fundamental of all principles"? Would John F. Kennedy have thought so? Harry Truman? Did Mondale believe this fundamental principle applied to South Africa?

Robert Kaiser, an editor of the *Washington Post*, described the administration's foreign policy as a "recipe for disaster":

> The invasion of Grenada will complicate the chances of any negotiated solution to the wars in El Salvador and Nicaragua. Mexico is already leading the chorus of Latin

American critics of the invasion. Their strongly negative reaction grows from an old and genuine anxiety about Yankee interventionism. It is now much less likely [it was never likely] that the so-called Contadora group of Central American states will be able to help the United States extricate itself from those two wars through negotiations.[58]

Mexico could always be relied upon to lead a chorus of criticism of the United States. As for the effect on Nicaragua, it was the opposite of what Kaiser predicted. Nicaragua immediately attempted to initiate talks with the U.S. government about all matters of mutual concern including security issues.[59] Sometimes a dose of fear can have salutary effects.

Kaiser's comment on negotiations is also revealing. He acknowledged hoping that the Contadora process would "help the U.S. extricate itself from these two wars through negotiations." This was ever the liberal hope. And it was precisely because our adversaries so often knew that we negotiated in hopes of "extricating" ourselves that our bargaining power was diminished.

Liberal Democrat Ted Weiss of New York believed the invasion of Grenada to have been an impeachable offense. He introduced an impeachment resolution cosponsored by Democrats John Conyers of Michigan, Julian Dixon of Illinois, Mervyn Dymally of California, Henry Gonzalez of Texas, Mickey Leland of Texas, and Parren Mitchell of Maryland.

The contrast between the liberal and conservative, or broadly speaking, Democrat and Republican views of Central America in the 1980s could hardly have been more dramatic. Conservatives viewed the Communist coup in Grenada with alarm. Liberals either yawned or welcomed it outright. The staff of Democratic congressman Ron Dellums of California became notorious for its enthusiastic embrace of Communist regimes. Carlottia Scott, a Dellums aide, enjoyed a lively correspondence with Maurice Bishop before his death, some of which demonstrated her boss's relationship with Castro.

Ron as a political thinker is the best around and Fidel will verify that in no uncertain terms.... Ron had a long talk with Barb and me when we got to Havana and cried when he realized that we had been shouldering Grenada alone all this time. Like I said, he's really hooked on you and Grenada and doesn't want anything to happen to building the revo [sic] and making it strong. He really admires you as a person, and even more so as a leader with courage and foresight, principle and integrity. Believe me, he doesn't make that kind of statement often about anyone. The only other person that I know of that he expresses such admiration for is Fidel.[60]

Carlottia Scott became "political issues director" for the Democratic National Committee in 1999. The "Barb" referred to in the letter is Barbara Lee, who is now a member of Congress (and the only one to vote against responding to the September 11 attacks with military power). When the U.S. military seized documents in Grenada, they found a reference to her in the minutes of a politburo meeting: "Barbara Lee is here presently and has brought with her a report on the international airport done by Ron Dellums. They have requested that we look at the document and suggest any changes we deem necessary. They will be willing to make changes."[61] Also in those minutes was a notation explaining: "Airport will be used for Cuban and Soviet military."[62]

## El Salvador: "Death Squads" and Democracy

For many (though not all) liberals, the fight in El Salvador was about poverty and misery and the desperation of people who were at the mercy of "right-wing death squads." There was violence on both sides—but that of the FMLN, the Communist insurgency, was regarded as a liberation force by the Left, and therefore excused. Liberals tended to assert that the Reagan administration was "supporting" right-wing violence, but this was false. In fact, the Reagan administration, and Assistant Secretary of State Elliott Abrams in particular, successfully pressured the Salvadorans to police their ranks. (Abrams had similar success in pressuring Chile to democratize.)

But conservatives, unlike liberals, understood the danger of an FMLN victory. A Communist El Salvador would introduce a level of misery, human suffering, and oppression to that country that would dwarf what the generals had imposed, as well as pose strategic problems for the U.S.

In the early part of Reagan's first term, El Salvador dominated the bitter and acrimonious debate. The Left was electrified, and groups sprang up all over the United States to support the Communist guerrillas in El Salvador. Committees in Solidarity with the People of El Salvador (CISPES), Americas Watch, the Washington Office on Latin America, Witness for Peace, and many other groups paid lopsided attention to the human rights abuses of right-wing Salvadorans and urged the view that only a revolution could save the country. Hundreds of churches offered "sanctuary" to Salvadorans who fled to America. It's a "moral response to an immoral foreign policy," explained one veteran of Vietnam War protests.[63] The movement did not offer sanctuary to Nicaraguans fleeing the Sandinistas.

The liberal press was unanimous that the U.S. policy of thwarting the Communist insurgency in El Salvador while simultaneously pushing for elections and a democratic government was doomed. And they certainly skewed reporting in a way that would increase the likelihood of failure. When the Media Research Center examined coverage of El Salvador in the major dailies, newsmagazines, and television news programs they noticed that during a one-month period of heavy coverage in 1989, the press never once labeled the FMLN as Communist. They were most often called "leftist," (123 times), "Marxist-led" (20 times), "left-wing" (12 times), or "Marxist" (5 times).[64] The Vietnam analogy was floated again and again. The very first question asked of President Reagan at his first press conference in 1981 was, "How do you intend to avoid having El Salvador turn into a Vietnam for this country?" And when Walter Cronkite interviewed the president, he asked, "Do you see any parallel in our committing military advisers and military assistance to El Salvador and the early stages of our involvement in Vietnam?"[65] Former U.S. ambassador to El Salvador Murat Williams said:

> We are blind to believe that supplying helicopters and weapons and Green Beret training is the way to solve the problems in El Salvador. You just don't solve social problems by military means. You can't keep the lid on the boiling cauldron of popular aspiration by using machetes and guns. Our government has been unwilling to recognize this fact. A lot has to do with the basic philosophy of the Reagan administration, which is that we are in a confrontation with the Soviet Union. It is all wrong to let our policies be determined by fear.[66]

In point of fact, it was the Carter administration that cut off aid to the Sandinista regime in 1980 due to its support for the FMLN guerrillas in El Salvador. Carter simultaneously resumed aid to the Salvadoran government.[67] But it was easy for liberals to forget those facts when Reagan's anticommunism made them dizzy with outrage. Also, the near universal assumption on the Left was that the existence of a Communist insurgency was evidence of the desperation and poverty of the people. And time and again, American policymakers were warned of the folly of attempting to "keep the lid" on such popular discontent. Yet never in history was a Communist regime *voted* into office. Further, it is revealing that those on the left rarely if ever argued that the answer to poverty and misery was free enterprise, free trade, and political pluralism. As Jack Kemp put it, "The Democrats were not soft on Communism, they were soft on democracy." (Well, both actually.) Nor did those who spoke so movingly of the plight of the poor in Third World nations ever express any comparable concern for the poor of Communist countries—so many of whom were desperate enough to flee their homes.

The war in El Salvador was vicious on both sides. Right-wing "death squads" did commit atrocities—though no more than the left-wing terrorists inflicted. (Communist killers were never identified as "death squads.") The war was misrepresented again and again by the American press and liberal politicians, who suggested that the Communist insurgency could not be defeated until the underlying conditions—poverty, illiteracy, hopelessness—that supposedly fed the revolutionary sentiment had been eradicated. (Twenty years on, liberals would argue that terrorism could not be defeated until its underlying causes—poverty, illiteracy, hopelessness—were cured.) Colorado Democrat Gary Hart offered that "the real problem of revolution is not communism, it's poverty." Some American liberals went even further, suggesting that the communists might actually be on "the right side of history" while the United States and its noncommunist allies were on the wrong side. Jesse Jackson, for example, said:

> Our embracing the landed gentry of the banana republic, our embracing of the Somoza regime—we've helped create a mess. We're about to make the same mistake of embracing the wrong side of history.... Nicaragua this past week made a judgment, to call for an election in November. We ought to recognize that government right now. We ought not to embrace El Salvador's killer regime.[68]

Maryland Democrat Barbara Mikulski was not far from Jesse Jackson. As a House member she bitterly denounced any U.S. involvement in Central America: "So here the United States is in Central America repeating the mistakes of Vietnam and Iran. Once again some of its officials are lying about terrorism and exploitation that the United States supports and helps pay for."[69] Massachusetts Democrat Barney Frank pronounced, "The administration would have us believe that, by sending arms, we are promoting peace and stability in Salvador. But the opposite is unfortunately true."[70] Michigan Democrat David Bonior warned:

> We are headed for disaster in this region. And if the American people and this Congress do not wake up to the fact that this nation is headed for a catastrophe very soon, a catastrophe that may involve the use of American troops, we are going to find ourselves—and I hate to use the parallel, because I think in many instances it is not accurate—but we are going to find ourselves in another situation that closely parallels that of Southeast Asia.[71]

Senator Edward Kennedy, the Massachusetts Democrat, opposed aiding the right-leaning government of El Salvador and helping it fend off the Communist assault:

> Rather than move toward a peaceful political solution in El Salvador, the Reagan administration has continued to pursue a policy of military escalation. Rather than work with our friends and allies in the region to resolve outstanding disputes within and among the nations of Central America, this administration has paid lip service to the idea of such cooperation, while attaching unacceptable preconditions to any negotiations.[72]

Like other Democrats, Senator Kennedy presented the issue as a contest between peace and war rather than choosing sides between communists and noncommunists.

Congresswoman Mary Rose Oakar of Ohio gave a flavor of the Democrats' debating tactic—impugning your adversary's motives and morality:

> One of the greatest sins during the Nazi era was silence. This administration has been silent about the thousands killed in El Salvador, silent about the deaths of the American

missionaries and the other Americans who were killed
needlessly, silent about the massacre in Lebanon. By their
silence they give the green light to massacres of people
throughout the world.[73]

On another occasion, Oakar demonstrated her familiarity with the
"lessons" of Vietnam, at least according to liberal Democrats: "Will
we continue to be the only major nation in our hemisphere that has
not recognized that guerrillas do not lose their homegrown democratic revolutions?"[74] There was nothing democratic about the FMLN.
Nor was it strictly "homegrown." It received copious assistance from
Cuba, Nicaragua, and the USSR. Finally, it was the Reagan policy
that ultimately produced a centrist government in El Salvador.

New York congressman Steven Solarz was considered one of the
moderates in the Democratic Party. But here is how he analyzed the
Reagan first term: "In almost every corner of the globe, four years of
Republican foreign policy have made the world a more dangerous
place to live. In Central America, where the administration has sought
military solutions to what are essentially political problems, its policies are leading us ineluctably toward the introduction of American
combat forces into El Salvador, and possibly Nicaragua as well."[75]

The Reagan foreign policy was certainly not perfect, but it was
nearly the perfect opposite of what Solarz described. In fact, around
the globe, President Reagan's policies pushed the world toward
greater peace and democracy. Among nations essentially friendly to
the United States, like Chile, El Salvador, and the Philippines, pressure
from the U.S. led to free elections and peaceful resolution of conflicts.
Nations hostile to America like the Soviet Union, Cuba, and
Nicaragua were forced to trim their ambitions in the face of American resolve. And nations like Afghanistan got a chance to repel a
Communist invasion. The United States never did send its own armed
forces into Central America. It was not necessary when there were
determined Nicaraguans and Salvadorans willing to fight for their
own freedom. The constant refrain from opponents of Reagan's Cen-

tral America policies—that providing military aid would make American military engagement more likely—were proved exactly wrong. If no military aid had gone to the contras and the Salvadoran government and those nations had become Soviet assets, American military engagement would have been more, not less, likely.

As with each and every Communist coup or attempted insurgency around the globe, many liberals declined to believe that the FMLN of El Salvador were really Communists. Appearing on *Meet the Press* in 1983, Senator Christopher Dodd, a Connecticut Democrat, was asked by Georgie Anne Geyer about his endorsement of negotiating with the Salvadoran guerrillas who were "avowed Marxists."

Dodd interjected, "Some of them."

Miss Geyer pointed out that the FMLN itself claimed to be Communist. Dodd was unconvinced. "Not all of them, though."

Geyer then asked, "Since these kinds of Marxists have never negotiated throughout history, except once in Czechoslovakia and only for a very short time...do you actually believe they would negotiate in good faith now?

Dodd replied, "...I don't know the good faith—no one knows that. You can only find that out by trying, sitting down and seeing if they're sincere, and testing it out....I don't know them. I know that they're not all Marxists, any more than all the Sandinistas were Marxists...."

Geyer came at him one last time on the nature of Communists: "Could you see that these—again, self-proclaimed Marxists—they would actually take part in a democratic process and observe the fruits and the outcomes of that process?"

Dodd was dogged. "They might. I don't believe Marxism is necessarily monolithic either.... We shouldn't assume that if someone happens to be Marxist that immediately they're going to be antagonistic to our interests or going to threaten our security."[76]

Congressman George Brown was equally nonchalant. Regarding the Marxist regime in Nicaragua, the California Democrat said, "There are Marxists, including Communists, in the governments of many of our allies. Some of these governments are our friends—one even controlled by the socialists."[77] Brown apparently saw little difference between socialists, who have often come to power through free elections, and communists, who never have.

Democratic senator Alan Cranston of California believed that:

> In country after country in Latin America, we have been indifferent to the poverty. We have backed tyrants or even imposed tyrants on the people.... Mexico tells us "Stop doing what you're doing in El Salvador. Stop backing the tyranny there. Stop trying to overthrow the government of Nicaragua. And we will then be able to take care of our problems."[78]

Many mainline churches sponsored "study tours" of Central America. The Unitarian Universalist Service Committee took members of Congress to visit "rebels and other dissidents not on the agenda of officially sponsored tours." Democrat congressman Jerry Studds of Massachusetts credited a Unitarian trip with making him one of the most vociferous opponents of aid to the Salvadoran government.[79]

When Democrat congressman George Crockett of Michigan sued President Reagan over aid to El Salvador, he was supported in his suit by the National Council of Churches, the United Presbyterian Church, the Church of the Brethren, the Unitarian Universalist Association, and others.[80]

The American Catholic bishops weighed in with Congress too, urging that aid to El Salvador be cut off—this despite the support for such aid from the Salvadoran Catholic bishops. Thomas E. Quigley, Latin American specialist at the U.S. Catholic Conference, declared, "I think many in the church would ... say that the ... FMLN are the embodiment of the legitimate aspirations of El Salvadoran people."[81]

He was probably right—many in the church would have said that—but they were all among the elites; they did not speak for the laity. One of the most notorious communist-sympathizing religious groups was the Catholic Maryknoll order, which enthusiastically endorsed liberation theology and even communism. In one of its publications, Maryknoll declared:

> For any Christian to claim to be anti-Communist, without a doubt constitutes the greatest scandal of our century. . . . For Jesus, whether conservatives like it or not, was in fact a communist. . . . Jesus explicitly approves and defends the use of violence. . . .[82]

The Maryknoll order exerted a disproportionate influence on U.S. policy because Speaker of the House Thomas P. ("Tip") O'Neill Jr., a Massachusetts Democrat, had an elderly aunt who was a member. She apparently gave him an earful of her views about El Salvador, and this convinced him to oppose Reagan's policies on Central America vehemently. His aunt, he told reporters, was his "secret source" of information.[83] The *New York Times* reported that "While some members of Congress base their foreign policy positions on elaborate briefings by aides, consultation with colleagues, or public opinion polling of their constituents, Mr. O'Neill depends upon the activist nuns and priests to help shape his views on Central America. . . . 'I have great trust in that order,' O'Neill stated. 'When the nuns and priests come through, I ask them questions about their feelings, what they see, who the enemy is, and I'm sure I get the truth.'"[84]

It showed. In 1984 O'Neill denounced the resistance in Nicaragua as "terrorists, marauders, and paid mercenaries." Alan Cranston thundered that, "Under Ronald Reagan, America's vision of world of peace and freedom is being blasted by the guns of the U.S. Navy off the coast of Lebanon, by the guns of the U.S. paratroopers in Grenada, and by the guns of U.S. helicopters in Honduras and El Salvador."[85]

Some Democrats were convinced that whatever the United States did must be for the worst motives. Congressman George Miller, a California Democrat, described America's El Salvador policy this way:

Institutionalized murder by the government of El Salvador is made possible by the dollars given by this administration with the consent of the Congress. The very survival of the oligarchy and the military corruption depends on the existence of the death squads, on political repression, and on American dollars.... We have made El Salvador into a client state and now we are a captive of that client.[86]

Democratic congressman Robert Garcia of New York wondered, "Who gives us the right to tell the people of Central America what to do.... It is not ours to win or lose. Central America belongs to the Central Americans.... It is important that we do debate this issue because today we may just be starting the next world war."[87]

In 1983, the U.S. conducted joint military maneuvers with the army of Honduras. Democratic congressman David Bonior of Michigan was convinced that this represented provocation and aggression by the United States:

By what principles do we justify these actions, and to what ends? Many of the consequences of these actions are already painfully clear. The uprooted, the wounded and the dead in Nicaragua bear witness to the human cost of the increasing militarization of the border region. The growing climate of fear in that country has led to increasing restrictions of freedom by a government that points to the threat of imminent war, and that may soon see itself with few options but to turn for protection from U.S. intervention to the forces that we would like least to see have a foothold in the region.[88]

This was January 1983. The Sandinistas had already taken huge strides toward turning their nation into a member in good standing of the Comintern. But as Jeane Kirkpatrick would label it at the Republican Convention eighteen months later, the reflexive response of Democrats was to "blame America first."

Democrat congressman (now senator) Byron Dorgan of North Dakota seemed to believe that the great threat to world peace was not Communism, but Ronald Reagan. In any case, he could see no moral difference between the free world and the Communist one:

> I have been to that region...and there is a claim that Nicaragua exports revolution. There is some small evidence of that. There are other claims that the United States imports revolution into Nicaragua, and there is some evidence of that. Under the guise of safeguarding our national security, Ronald Reagan has made of the United States an outlaw nation in the eyes of much of the world.... Whether Ronald Reagan is capable of accepting it or not, *state-sponsored terrorism is reprehensible whether committed by us or by the Russians* [emphasis added]. Ronald Reagan has once again managed to call into worldwide question his balance and judgment when it comes to matters of peace and war.[89]

Dorgan's comments were tendentious beyond belief. Note the lack of the modifier "small" when he spoke of the evidence that the U.S. was importing revolution to Nicaragua and its presence when referring to evidence of Sandinista support for El Salvador's FMLN. Dorgan drew no distinction between those attempting to export tyranny and those attempting to import freedom—not a small point, one would think. Also, for whom was he speaking when he asserted that Reagan had made the U.S. an "outlaw" nation? The only countries who spoke that way about the U.S. were our enemies. Certainly the free nations of Europe were in the process, even as Dorgan spoke, of following American leadership on the matter of intermediate range nuclear missiles. As for the accusation that America was a state sponsor of terror because it supported the noncommunists—such a shameful accusation could easily have come from the pages of *Pravda*.

In time, the policy of the Reagan administration—to push for democratic reforms while at the same time countering the communist insurgency with force—was successful. In 1984, despite threats and intimidation from the FMLN, millions of Salvadorans waited in long lines to vote. The man they elected, Christian Democrat Jose Napoleon Duarte, had been imprisoned and tortured by a previous right-wing government and thus his bona fides could not be challenged.

After Duarte's election, El Salvador retreated from the headlines. But its place in the heated debate over the Cold War was immediately taken by Nicaragua.

## NICARAGUA: UNPACK THE CLICHÉS

As they did with every other nation that had ever been overtaken by Communist guerrillas, liberals at first believed the worst about the Somoza regime. The *New York Times* editorialized: "The tyrant we know in Nicaragua...seems worse than any possible successor....If North Americans show themselves unafraid of [Somoza's] red scare stories, we may yet help to bring better times to Nicaragua."[90] In an almost eerie repetition of its early stupidity about the Khmer Rouge, the *New York Times* advised its readers that the FSLN was a "broad-based organization that was founded in 1962 and now includes Roman Catholic youths, young businessmen, peasants, priests, and leftist students."[91]

Though it would later be alleged (again) by liberals that American hostility drove the Sandinista regime into the arms of the Cubans and Soviets, the truth is that the United States helped to topple Somoza in the first place (imposing an arms embargo in 1978). With full U.S. participation, the Organization of American States, for the first time in its history, issued a call for the replacement of the Somoza regime with a pluralistic, democratic opposition. The Sandinistas pledged to implement the reforms demanded by the OAS, including a promise to hold early elections. Taking them at their word, the U.S. offered generous aid and friendly relations with the new Sandinista Directorate when it first took power. The Carter administration provided $118 million in direct aid, negotiated with the international development banks for an additional $262 million, and arranged for refinancing of $500 million in private bank debt.[92] This was more aid than any other nation provided to Nicaragua, and more than the U.S. had contributed to Somoza in the previous four years.[93] Jimmy Carter also welcomed Daniel Ortega to the White House.

Though the record left little doubt about the Sandinistas' ideological leanings, the U.S. State Department and the Carter administration remained agnostic about its true character. They pointed to the noncommunist members of the Sandinista Directorate and grasped at other straws. Deputy Secretary of State Warren Christopher (later secretary of state in the Clinton administration) testified before a House committee in September 1979 that "The [Nicaraguan] government's orientation, as revealed in its initial policies, has been generally moderate and pluralistic and not Marxist or Cuban." Christopher further assured the Congress that the Sandinistas were desirous of "close and friendly relations" with the United States.[94] Even after the new leaders had banned the opposition, begun to militarize the country, and visited Cuba, President Carter resented the suggestion that this should alarm us:

> It's a mistake for Americans to assume or to claim that every time an evolutionary change takes place in this hemisphere that somehow it's a result of secret, massive Cuban intervention. The fact in Nicaragua is that the Somoza regime lost the confidence of the people. To bring about an orderly transition there, our effort was to let the people of Nicaragua ultimately make the decision on who would be their leader—what form of government they should have.[95]

Declining to fear inordinately—or perhaps even to dislike—communism was President Carter's North Star. (This was several months before Soviet tanks rolled into Afghanistan.) Carter was still battling Joe McCarthy. But the reality in 1979 was that Cuba was openly, not secretly, aiding the Sandinistas, though, to be sure, Castro advised the Nicaraguans to be cagey about their ideology. One of the "commandantes" said later "Our strategic allies [i.e. Cuba] tell us not to declare ourselves Marxist-Leninist.... [Nicaragua] will be the first experience of building socialism with the dollars of capitalism." Still, close observers could not miss the radical character of the Sandinista regime, and those who were fooled by the subterfuge were eager to be fooled.

Among those was foreign correspondent Karen de Young of the *Washington Post*, who described the Sandinitas as "hazy in ideology." The *New York Times* couldn't quite make out who the Sandinistas were either, calling them "politically ill-defined."[96]

Though they lied about it, the Sandinistas began funneling money to the FMLN in El Salvador from their first days in power. In 1981, Tomas Borge proclaimed, "This revolution goes beyond our borders." That same year a leader of the FMLN in El Salvador said, "We have the brilliant example of Nicaragua.... The struggle in El Salvador is very advanced; the same in Guatemala, and Honduras is developing quickly.... Very soon Central America will be one revolutionary entity."[97] Pace Jimmy Carter, the Sandinistas were the last people to permit the Nicaraguans to make "the decision about who would be their leader."

It is difficult to understand how American liberals, having watched Communist regimes repeat the same pattern again and again for sixty years, could have failed to anticipate that the Sandinistas, once in power, would squeeze out all opposition, shut down the free press, and militarize the country. Yet time after time, liberals would follow their own well-worn pattern: first deny that the revolutionaries were Communists; second, assert that the no longer deniable repression was a response to U.S. actions; and third, laud the accomplishments of the revolution while offering perfunctory criticism (if that) of the human rights situation.

Then representative (now senator) Tom Harkin, an Iowa Democrat, was typical. In 1979 he analyzed the Nicaraguan situation as follows:

> Nicaragua is a very small country. Thus, the members of the Broad Opposition Front are quite familiar with the members of the provisional government and would know, much better than we would, if there was reason to be wary of it. The fact that the Broad Opposition Front, whose members would have the most to lose by a Communist takeover, would endorse the new plan, should assuage fears in this country that the junta is dominated by Communists.... The chances that General Somoza will be replaced by a truly democratic and representative government can only be enhanced by this most recent show of unity.[98]

While President Carter and the U.S. State Department attempted to woo the Sandinistas with aid and fond hopes for good relations, the Nicaraguans were setting their faces firmly the other way. A record of Sandinista deliberations in 1979 was later smuggled out of the country. Called the "Seventy-two-hour Document," it revealed the classic Marxist terms in which the Sandinistas referred to themselves. The Frente Sandinista de Liberación Nacional (FSLN) was the "vanguard" of the revolution, it said. Its alliance with more democratic anti-Somoza groups was merely tactical—an attempt to "neutralize" domestic rivals and ward off "Yankee intervention." In fact, once in power, the Sandinistas intended to "crush" internal opposition and aid revolutionary movements abroad.[99] Humberto Ortega, Nicaraguan defense minister and brother of Daniel, the Sandinista president, affirmed that "our revolution has a profoundly anti-imperialist character, profoundly revolutionary, profoundly classist; we are anti-Yankee, we are against the bourgeoisie... we are guided by the scientific doctrine of the revolution, by Marxism-Leninism."[100] On the second anniversary of the revolution, Tomas Borge, minister of the interior and head of the secret police, proclaimed, "Our revolution was always internationalist from the moment Sandino fought in La Segovia."

The Sandinistas also undertook other, familiar, Communist measures. They closed several radio stations and put them under new ownership as Radio Sandino. The nation's two television stations were incorporated into the new government-controlled Sandinista Television System. And the renowned newspaper *La Prensa*, which had played a critical role in unseating Somoza, was strictly censored and frequently harassed.

As in every other communist country, the Sandinista regime intimidated, harassed, and tortured labor leaders. The AFL-CIO, a staunch anticommunist union throughout its history, shamed its liberal friends by its clarity about Nicaragua. Lane Kirkland, head of the AFL-CIO, said, "Nicaragua's headlong rush into the totalitarian camp cannot be denied by any who has eyes to see."[101] But few liberals and no leftists had eyes to see. One liberal organ that did depart from the usual script was the *New Republic*. Always an eclectic blend, the magazine was responsible for some of the most influential anticommunist essays published in the 1970s and 1980s. Editor Martin Peretz oversaw a series of articles and editorials, including a highly controversial endorsement of aid to the Contras, that dismayed many liberal supporters of the magazine. Professor Stanley Hoffman of Harvard wrote a letter to the editor denying that Nicaragua was being used as a Soviet base, and lamenting that "once again, we are being served an amazingly angelic view of our own side, an entirely scary one of 'the enemy,' and a cheerfully unrealistic assessment of what the proposed course would achieve."[102] Another letter, signed by many contributors to the *New Republic*, urged "political and economic pressure, negotiations, and aid and encouragement to democracies in the region." The writers were convinced that "even with massive American aid the Contra army will be militarily and politically unable to succeed..." and that anyway, it didn't deserve to since it was "dominated by antidemocratic supporters of the former Somoza regime."[103] The letter was signed by Abraham Blumberg, Robert Coles, Henry Fairlie, Hendrik Hertzberg, Vint Lawrence, R. W. B. Lewis, Mark Crispin Miller, Robert B. Reich, Ronald Steel, Richard Strout, Anne Tyler, Michael Walzer, and C. Vann Woodward. As it turned out, the Contras succeeded beyond all expectations militarily and especially politically.

The Sandinistas had enjoyed good relations with the Palestine Liberation Organization for a decade. Sandinista militants had received training in Fatah camps in Lebanon and Libya during the 1970s and had fought with the PLO against Jordan's King Hussein. In 1970, Sandinista Patrick Arguello Ryan was killed while participating in the hijacking of an El Al jet from Tel Aviv to London. The Sandinistas named a large dam after him.[104] Nicaragua's new rulers made their anti-Jewish and anti-Israel views known from the beginning and quickly moved against Managua's small Jewish community. In 1978, Sandinista gangsters had set fire to the doors of Managua's synagogue while a service was in progress. Those who attempted to exit through a side door were forced back by armed men. The fire was extinguished and there were no deaths. After the revolution, the Sandinistas confiscated the synagogue and Jews received death threats and other harassment. One tactic was to phone people in the middle of the night to tell them that their relatives had been shot and killed. Businesses were expropriated, and trumped up charges brought against individual Jews. The president of Managua's synagogue was arrested and forced to sweep the streets. His textile factory was confiscated. Anti-Semitic slogans ("Death to Jewish Pigs") and graffiti marred the synagogue while the Sandinistas welcomed the PLO warmly. The fifty families comprising the Nicaraguan Jewish community, who had lived in Managua for more than one hundred years, fled.

The Catholic Church had been instrumental in the removal of Anastasio Somoza, but the Sandinistas wasted little time in attacking the Church once the revolution was over. The government launched a campaign of vilification against the archbishop, Obando y Bravo, calling him a priest for the rich, and insisted that Catholic schools teach Marxism-Leninism. The Sandinistas also began to draft seminarians into the army and forbade the broadcast of church services on radio. Obando y Bravo was forbidden to perform his popular outdoor masses, and other church leaders were repeatedly arrested and interrogated by state security.

Protestants, too, were harassed. Fifty Moravian churches were burned. Sandinista mobs ("turbas divinas") were sent to disrupt a "Day of the Bible" celebration planned by several Protestant denominations, and visiting preachers were forbidden to address the crowd. Members of the Campus Crusade for Christ, the First Evangelical Church of Central America, and National Council of Nicaraguan Evangelical Pastors were stripped naked and held in cold, dark cells for hours before being interrogated.[105] Sandinista soldiers captured one itinerant Protestant missionary, Prudencio Baltodano. They beat him, slit his throat, and sliced off both of his ears before leaving him tied to a tree to die. But he survived and escaped to tell his story.[106]

The Moskito Indians were a Protestant group living on the Atlantic coast. They had a separate ethnic identity from other Nicaraguans, and most governments in Managua had left them alone. But the Sandinistas replaced their local teachers with Cubans and insisted that the children be taught Marxism-Leninism. When this led to a revolt, the Moskitos were ruthlessly attacked and forced from their homes. Several thousand were forcibly relocated to concentration camps in the interior. Their crops and churches were burned and their homes destroyed. Eden Pastora, a member of the original Sandinista leadership who later defected and led a resistance group, protested this atrocity: "Even the tyrant Somoza left them alone. He might have exploited them a bit, but you want to turn them into proletarians by force!" Tomas Borge, minister of the interior and head of state security replied, "The revolution can tolerate no exceptions."[107] This massive assault on a poor and defenseless people was somehow overlooked by the United Nations—and by nearly all the groups who styled themselves "human rights" organizations.

Even while receiving American aid during the first eighteen months of its life, the Sandinista regime rapidly militarized the country along Cuban and Soviet lines. By 1980, the Sandinista army had 12,000 men under arms, double the number of soldiers that Somoza had employed in the National Guard. The armed forces were beefed up to 75,000 by 1985—26,000 more than the next largest military in Central America, El Salvador's. Nicaragua, a nation of 2.8 million amassed an army larger than that of Mexico, with a population over 70 million. To facilitate this transformation of the Nicaraguan military, the Sandinistas welcomed the help of 3,000 Cuban and several hundred Soviet and Eastern European advisers.[108]

It would later be argued that Nicaragua had engaged in a military buildup in response to the threat from the U.S.-backed Contras, but as their own documents reveal, the arms buildup began in 1979, long before the first Contra took up a rifle and while the U.S. was still providing aid. The Soviets were lavish with weapons, providing 350 T-55 tanks, 12 MI-8 HIP helicopters, six HIND attack helicopters (the weapon that so devastated Afghanistan), as well as artillery pieces, patrol boats, and airplanes. With Soviet and Cuban help, the Nicaraguans built a military air base at Puenta Hueta, with a runway 10,000 feet long—long enough to accommodate any plane in the Soviet arsenal.[109]

Like other Communist states, Nicaragua found itself in urgent need of more prisons and quickly built them. By 1983, in addition to the estimated 4,000 former members of Somoza's National Guard, Nicaragua held 6,500 political prisoners, more than any country in the western hemisphere except Cuba and a remarkable number for a nation of only 2.8 million.[110] As in Cuba, political prisoners were kept in execrable conditions, sometimes with hands tied above their heads for many hours, or in solitary confinement. The Sandinista innovation was the tiny cell, too small even to sit down in. These "chiquitas" were completely dark, had little air, and no sanitation. Prisoners were kept in these torture chambers for up to a week.[111]

Like all other Communist regimes before it, the Sandinistas also produced the one, most reliable product of communism—refugees. Ninety-five thousand Nicaraguans fled to Costa Rica following the Sandinista coup in 1979, while thousands of others made their way to Honduras and to the United States. About one in ten Nicaraguans fled their homes in the face of Sandinista repression.[112]

This then was the true face of communist Nicaragua—a repressive, totalitarian tyranny like its many predecessors—committing crimes against its own people, attempting to subvert the governments of its neighbors, a client of Cuba and the Soviet Union, and a friend of terrorists.

None of these attributes gave liberal and leftist fans of the Nicaraguan regime any pause. For the hard Left, Nicaragua became a hero state as soon as the Sandinistas took power. The usual political pilgrims—Susan Sontag, William Sloane Coffin, Richard Falk, Pete Seeger, Benjamin Spock, Noam Chomsky, Jessica Mitford, and many more hopped the nearest plane to breathe in the invigorating revolutionary air in Managua. The "Sandalistas," as they came to be called, were greeted handsomely by the nine "commandantes" of the Sandinista Directorate. Mindful of the Vietnam example (where, they believed, American public opinion had undermined the war effort), the Sandinistas spent a good deal of time and treasure courting influential Americans. A correspondent for *Playboy* magazine recounted: "Here was a place seemingly run by the kind of people who were Sixties radicals. Wherever we went, people were young, singing political folksongs and chanting 'Power to the People.' One night there was even a Pete Seeger concert in town."[113] Of course there was.

James C. Harrington, the legal director the Texas Civil Liberties Union (a chapter of the ACLU) gushed about his trip: "We met Sergio Ramirez, two department directors...Vice-Foreign Minister Nora Astorga (a charming heroine of the revolution)...the Minister of Culture (Father Ernesto Cardinal) and...two of the three Electoral commission members."[114] Harrington did not say whether the members of the electoral commission had anything to do all day, since Nicaragua did not conduct elections. [In 1984, under strong international pressure, the Sandinistas did consent to an election, but it was a sham and was not recognized as free by any neutral observer.] As for Nora Astorga, her "charming" contribution to the revolution was to lure one of Somoza's generals into her bed so that five armed Sandinistas could jump from the closets and slit his throat. Ms. Astorga went on to serve as one of the Sandinistas' chief prosecutors, sentencing 7,500 Nicaraguans—anyone suspected of hostility to the new regime—to long prison sentences.

Mary Travers of Peter, Paul, and Mary traveled to Nicaragua as the "special guest" of President Daniel Ortega, and she threw parties for the Sandinistas here.[115] Betty Friedan and Abbie Hoffman made pilgrimages in the company of professors, politicians, lawyers, and activists. Actors Ed Asner, Kris Kristofferson, Michael Douglas, and Susan Anspach were enthusiastic, as were Jackson Browne, Mike Farrell, and Diane Ladd.[116] Managua became the favorite destination of an entire generation of radicals whose principal animating passion was hatred for their own country. Any nation, no matter how noxious, that opposed the U.S. had an automatic claim upon their sympathy.

In his memoir, former leftist Ron Radosh recounts his attempt to persuade members of *Dissent* magazine's editorial board to rethink their support for the Sandinistas. After documenting the Sandinistas' ghastly record on human rights and on links to the Soviet Union, Radosh sat back to hear what these New York intellectuals would say. One editor put into words what seemed to motivate so many: "You may be right about what you say about the Sandinistas," she said, "but while they are under attack by the American empire, we have a responsibility to extend our solidarity to them."[117]

A much watched celebrity in Managua during the early 1980s was Bianca Jagger, former wife of Rolling Stone Mick, who chatted up visiting journalists at the InterContinental Hotel in Managua. A native of Nicaragua, Jagger was a one-woman Chamber of Commerce for the Sandinistas. Radosh soon learned just how intimate her relations with the junta were when two or three evenings a week, Minister of the Interior Tomas Borge would arrive, surrounded by bodyguards, and head upstairs to Jagger's suite for the night.[118]

Many of the journalists covering Nicaragua were eager to give the Sandinistas the benefit of the doubt. On the occasion of a huge rally to inaugurate the new regime, the *Washington Post*'s Karen de Young reported:

> While the junta's program remained vague, today's inaugural and victory speeches contained little to sustain fears shared by the United States and conservative Nicaraguans that the country will move far to the left. Rather, the junta members and guerrilla leaders spoke of unity, the elimination of vestiges of Somoza and the hard task ahead of reconstructing a devastated country.[119]

Charles Krause, another *Post* reporter assigned to Nicaragua, showed a great deal more understanding of the nature of the Sandinista regime. Unfortunately, de Young was more typical of the press in general. Writing in the *New York Times* at about the same time, Alan Riding was all optimism: "For the former rebels, the consuming interest in their lives has switched overnight from destruction to construction. . . . The segments of the population that are feeling the revolution most keenly are the city slums. . . . The barrio, formerly a breeding ground for disease, crime, and despair, has today developed a new sense of purpose and unity."[120] Later in the same article (only weeks after the Sandinista victory), Riding claimed that the countryside was transformed as well: "The government worked quickly to form communes and give possession of the farms to landless peasants. Thousands of landless peasants are now working their 'own' land, and the cheerful mood of the countryside contrasts sharply with that of the cities."[121] Despite the slum clearance? Years later, the *New York Times* would run a correction which demonstrated that its confusion about Communism and the Sandinistas persisted to the bitter end: "A television review on November 27...about the Iran-Contra affair, described the Sandinista government of Nicaragua incorrectly. While Marxist, it was not Communist."[122]

Shirley Christian, one of the few reporters to remain unseduced, wrote:

> Reporters covering the war saw Somoza's opponents, the Sandinistas, through a romantic haze. This romantic view of the Sandinistas is by now acknowledged publicly or privately by virtually every American journalist who was in Nicaragua during the two big Sandinista offenses, the general strikes and the various popular uprisings. Probably not since Spain has there been a more open love affair between the foreign press and one of the belligerents in a civil war.[123]

The love affair engaged the affections of the broadcast media as well. On the eve of the Sandinista takeover, NBC correspondent Ike Seamans explained that "A number of Sandinista leaders are communist, but provisional government leader Moises Hassan says that does not necessarily mean Nicaragua will be communist." Next, Hassan was shown promising that "Our main goal now...is to bring democracy [so] that the people can express their will and choose what they want."[124]

While the press presented a mild view of the Sandinistas, plenty of Americans were outright cheerleaders for the new Communist on the block.

Michael Harrington, founder of the Democratic Socialists of America, returned from a trip to Nicaragua full of indignation:

> I came back from Nicaragua far more ashamed of my country than at any time since the Vietnam war. The Nicaraguans are a generous people, a poor and often

hungry people, who want to make a truly democratic revolution and it is we who work to subvert their decency."[125]

Ruth Harris, an official with the United Methodist Church Board
of Global Ministries, was of the same mind. After traveling to
Nicaragua with a delegation of other church leaders, she reported
feeling "humiliated and angry" at U.S. policies for the region which,
she said "dismayed us beyond words."[126]

Members of the U.S. Congress, too, were convinced that despite
its many human rights abuses and connivance with Cuba and the
Soviet Union, Nicaragua was engaged in a valuable experiment that
the United States should respect. As late as 1984, California Democrat Matthew Martinez defended the Sandinistas on the House floor:

> At first the revolution in Nicaragua was welcomed by lead-
> ers of this country. But when the Nicaraguan leaders
> looked beyond the American style of democracy, to estab-
> lish a unique, indigenous form of government tailored to
> accommodate the needs of that nation, the revolution was
> declared as having been lost to Soviet-Cuban ideology and
> subsequently targeted for U.S. supported subversion and
> terrorism. I say that this perception is wrong. Revolutions
> take time to mature...."[127]

Vermont Democrat Patrick Leahy took to the Senate floor with
almost equally muddled thinking: "I embarked on my fourth trip to
Central America in the last six years convinced the policies of the
Reagan administration in Central America are leading toward a major
foreign policy disaster for our country....It is clearer than ever that
the President's policy of military confrontation with Nicaragua is
sinking the United States deeper and deeper into a quagmire...."[128]

Democratic senator Claiborne Pell of Rhode Island said, "I believe
that the administration's policy of pressuring the government of
Nicaragua by one means or another will serve only to self-fulfill a
prophecy that Nicaragua is a Marxist regime, firmly in the Soviet

camp, and a carbon copy of Cuba. Yes, there are problems in Nicaragua, but they are being exacerbated by the pressures from outside, much of it from this country."[129]

Senator Ted Kennedy agreed, decrying a policy that "has helped the Sandinistas to consolidate support among the Nicaraguan people by identifying their revolution with Nicaraguan nationalism and has given the Sandinistas a pretext to pursue repressive policies within Nicaragua, thereby further entrenching their power within the country."[130]

Nor did it trouble liberal Democrats that the Sandinistas clearly had no intention of holding the free elections they had promised in 1979. In 1981 Humberto Ortega, Sandinista minister of defense, said, "Keep firmly in your minds that these elections are to consolidate revolutionary power, not to place it at stake." Sergio Ramirez, another member of the ruling junta, said, "The Nicaraguan people will have to choose and vote for one candidate. That candidate is the revolution. This is very important."[131]

Democratic congressman Michael Barnes of Maryland was certain that President Reagan's policies had provoked the Sandinistas and "pushed [them] farther away from the negotiating table and into the willing embrace of the Soviets."[132] And in perhaps the prototypical expression of naiveté, House Majority Leader (later Speaker) Jim Wright offered that, "Everyone with whom we talked believed...that a show of friendship by the United States would influence political developments for the better."[133]

At times, the Democrats' contempt for President Reagan and his anticommunist foreign policy led them to conduct that verged on disloyalty. In March of 1984, eight members of Congress signed a letter to Nicaraguan president Daniel Ortega. "Dear Commandante," they wrote, "We address this letter to you in a spirit of hopefulness and good will. As members of the U.S. House of Representatives, we regret the fact that better relations do not exist between the United States and your country."[134] The implication, obviously, was that the Reagan administration was to blame for the poor state of relations between the two nations. If only the administration possessed the "hopefulness and goodwill" of congressional Democrats, things would presumably be different. The letter next praised Ortega for *promising* (1) to hold free and fair elections, (2) to "reduce" press censorship, and (3) to permit "greater freedom of assembly for political parties."[135] They further praised Ortega for taking these "steps in the midst of ongoing military hostilities on the borders of Nicaragua." But the "steps" were not steps at all, merely promises. To be sure, the thrust of the letter was to urge the Sandinistas to make good on these pledges, but the supplicatory tone, along with the implied criticism of President Reagan's more resolute approach ("ongoing military hostilities") amounted to a coded signal to the Sandinistas that they had eager friends in the U.S. who were seeking, in the words of the letter, "to significantly strengthen the hands of those in our country who desire better relations...." The letter was signed by Democrats Jim Wright of Texas, Michael Barnes of Maryland, Bill Alexander of Arkansas, Matthew McHugh of New York, Robert Torricelli of New Jersey, Edward Boland of Massachusetts, Stephen Solarz of New York, David Obey of Wisconsin, Robert Garcia of New York, and Lee Hamilton of Indiana.

Aides to Senators Dodd and Harkin routinely met with Sandinista officials, reportedly coaching them on relations with the U.S. Congress.[136] Democratic congressman David Bonior's private contacts with Sandinista officials were intercepted by U.S. intelligence agencies.[137] And Senator Harkin actually traveled to Managua to urge Violetta Chamorro, editor of *La Prensa*, to accept limited press censorship.[138]

To call it naiveté is not quite apposite, because the Democrats who were convinced that friendship and negotiation were the keys to solving problems with Communist regimes and empires had no trouble morally condemning nations of the so-called "Right"—nations like South Africa, Chile, and El Salvador. It is impossible to imagine the eight congressmen who signed the "Dear Commandante" letter penning a similar entreaty to the leader of South Africa or El Salvador.

A study by the National Conservative Foundation found this inconsistency particularly evident in the press. Examining coverage of human rights abuses in 1984, the NCF monitored reports by the *New York Times*, the *Washington Post*, *Time*, and *Newsweek* on Central America. They found that these leading voices of the establishment devoted five times more coverage to stories about Salvadoran government human rights violations than to Sandinista offenses. While 216 stories in those publications concerned the deaths of four American nuns and two labor organizers by government forces, only seven stories reported the massacre of thousands of Moskito Indians by the Sandinistas. And while the American angle in the Salvador report (they were American nuns) perhaps accounts for some of the extra interest in that story, the prominence of Nicaragua in the national debate in Washington would certainly justify more than the single story *Newsweek* reported all year on Sandinista human rights abuses (this in contrast to fourteen stories about Salvadoran excesses).[139]

In 1988 the Media Research Center conducted a similar survey examining the coverage of television news. Monitoring the CBS *Evening News*, ABC's *World News Tonight*, NBC's *Nightly News*, and CNN's *PrimeNews*, MRC looked at how the television networks reported on Nicaragua. They found that Sandinista promises to live up to agreements were lavishly covered while truculent or defiant statements by Sandinista leaders tended to be ignored. In a May 1988 speech, Daniel Ortega mocked the "peace process," saying that the city of Managua would have to be "disinfected" following the departure of Contra negotiators. None of the four major networks aired a story about this. But when the Sandinistas promised, two months earlier, to offer amnesty to the Contras, permit freedom of speech, and release political prisoners in exchange for concessions from the United States, all four networks carried stories. Five times as many stories covered Sandinista reforms as mentioned Sandinista violations.

When Costa Rican president Oscar Arias issued a report itemizing Sandinista noncompliance with promised reforms, the story was buried. So, too, was President Daniel Ortega's vow to "crush the

Contras." When Sandinista mobs ("turbas divinas") attacked a peaceful march by mothers protesting the military draft, the networks ignored it.[140]

Democratic congressman David Bonior of Michigan, who presumably enjoyed access to news outlets other than television, acknowledged that the Sandinistas had made errors, "but if we measure the errors of Nicaragua, if we lay them side by side with what might be called the errors of our own administration, *which seem greater to me* [emphasis added], if we look at our conscience and assess the factors [at] play in Central America, who would we conclude poses the greatest threat to the stability of the region, Nicaragua or the present administration?"[141]

A delegation of celebrated writers certainly would have agreed with Bonior. Graham Greene, Günther Grass, Carlos Fuentes, Gabriel Garcia Marquez, Julio Cortazar, William Styron, and Heinrich Boell signed an open letter to the American people pleading that the U.S. not crush the "modest but profound achievements of the Nicaraguan revolution."[142] The Reverend William Sloane Coffin, (who should have earned a lifetime achievement award from the Kremlin), described the aims of the Sandinistas as "to stop the exploitation of the many by the few."[143] He also dismissed the notion that the Sandinistas were exporting revolution to El Salvador: "As for the charge that the Sandinistas are exporting violence.... Anyone with common sense knows that you can't have revolt without revolting conditions."[144] Leaving aside the speciousness of his point ("revolting conditions" have unfortunately been a feature of human societies since the beginning of time and hardly give rise in every case to communist revolutions) it is clear that for Coffin, the export of revolution to El Salvador would be no bad thing.

Coffin was not simply anti-anticommunist, as he claimed. All of his sympathy flowed left. When Nicaraguan president Daniel Ortega visited New York in 1986 (where he picked out expensive designer eyeglasses on Madison Avenue), he was a guest of honor at Coffin's

Riverside Church. The previous year, Coffin had offered this prayer from the pulpit:

> In repentance lies our hope....Were we Americans to repent of the self-righteousness...we would realize that if we are not yet one with the Soviets in love, at least we are one with them in sin....Were we to repent of our self-righteousness, the existence of Soviet missiles would remind us of nothing so much as our own. Soviet threats to rebellious Poles would call to mind American threats to the Sandinistas. Afghanistan would suggest Vietnam. Soviet repression of civil liberties at home would remind us of our own complicity in the repression of these same civil rights abroad....Jesus would never be "soft on communism" anymore than He would be soft on capitalism.[145]

The Reagan administration's policy toward Nicaragua evolved slowly. One early step, the secret mining of Nicaragua's harbors, was less than successful. The mines were so-called "firecracker" mines that did no real damage to shipping but were apparently intended to harass nations trading with the Sandinistas. When the mining became public, members of Congress rushed to microphones to proclaim their horror (though some of those protesting most grievously had been briefed on the operation in executive sessions). For all the furor their discovery provoked, they might as well have been genuine mines.

President Reagan had also authorized funding for a small force of Nicaraguans to take up arms against the Sandinistas. "Contras," cried the Sandinistas, meaning "counterrevolutionaries." But the label was never accurate. Most of those who fought against the Sandinistas were either former Sandinistas themselves or ordinary peasants who had good and sufficient reasons for hating and fearing them. (The peasants in the USSR, too, had taken up arms against the Communist government in the early 1920s—and they were massacred by the millions through starvation, relocation, and simple bullets.)

Whether to aid anticommunist insurgencies around the world, but particularly in Nicaragua, became one of the most contentious issues of the decade—revealing the deep chasm between liberals and conservatives.

In a prime-time televised 1983 speech, President Reagan made the case for aid to the Contras. [The term *Contra* is used here not in the spirit the Sandinistas used it, but rather, to mean contra-Sandinista.] The response from liberals and leftists was emphatically negative.

Democrat congressman (eventually senator) Charles Schumer of New York said:

> It seems that from his speech last night, the president came out of a time warp. He arrived from 1967 and came to us today, with different names and different places, but the same exact arguments that we were hearing in the middle sixties about Vietnam.[146]

Connecticut senator Christopher Dodd delivered the televised response for the Democratic Party. He warned that the U.S. was in danger of standing against "the tide of history"—an extraordinary statement under any circumstances but particularly foolish in light of the sudden collapse of Communism that was only seven years off at that point. Unsurprisingly, Dodd diagnosed the problem in Central America as poverty. "If Central America were not racked with poverty, there would be no revolution." And he was convinced that aiding the Salvadoran government would only make things worse. "American dollars alone cannot buy military victory [though Dodd elsewhere argued the military victory was impossible in any case]—that is the lesson of the painful past and of this newest conflict in Central America. If we continue down that road, if we continue to ally ourselves with repression, we will not only deny our own most basic values, we will also find ourselves once again on the losing side...."[147]

As for the nature of the Nicaraguan Contras, Dodd repeated the libels fashionable on the Left (many of which had originated in Sandinista propaganda). "The insurgents we have supported are the remnants of the old Somoza regime—a regime whose corruption, graft, torture, and despotism made it universally despised in Nicaragua."[148]

Actually, while the Contra leadership in its earliest days did contain a large number of former National Guard officers, the composition of the force changed very rapidly thereafter. One of the three leaders of the movement, Alfonso Robello, was a former member of the Sandinista junta. Arturo Cruz and Adolfo Calero were both opponents of Somoza, and Calero was imprisoned for his anti-Somoza agitation. Among the military leadership of the Contras, 53 percent were former civilians, 27 percent former guardsmen, and 20 percent former Sandinistas. As the *New Republic* editorialized, "One doesn't raise an army of 15,000 peasants with promises of restoring a universally despised dictatorship."[149]

Heavy hitters from the Democratic foreign policy establishment leant their weight to the anti-contra cause. McGeorge Bundy, national security advisor to Presidents Kennedy and Johnson; Clark Clifford, secretary of defense to President Johnson; Robert McNamara, secretary of defense to Presidents Kennedy and Johnson; Edmund Muskie, former senator from Maine and secretary of state for President Carter; and Cyrus Vance, President Carter's first secretary of state, all signed a mailgram to members of Congress urging defeat of contra aid. Liberal Republican Elliott Richardson joined them.[150]

Liberals argued that the Contras were mere mercenaries bought and paid for by the CIA. While it was true that the CIA provided initial seed money for the force, the growth of the Contra movement was not dependent upon U.S. aid. Peasants from the countryside, Moskito Indians, and other desperate Nicaraguans swelled the Contra army to seven thousand men by 1983—more than the Sandinistas had under arms when they toppled Somoza. During the period when the U.S. provided no assistance (May 1984 to August 1985), the Contra army grew by another six thousand volunteers.

By 1987, the Contra army was twenty thousand strong and putting painful pressure on the Managua regime and its army of conscripts.

While liberals blasted the Contras for being CIA tools, they never asked where the Salvadoran FMLN got its backing. In fact, they argued again and again that the FMLN was purely homegrown and indigenous, despite its lavish backing by Cuba, Nicaragua, and the Soviet Union. Yet the Contras were regarded as tainted because they had received CIA backing.

Liberals also seemed to think that because the United States had sometimes backed dictatorial regimes in Latin America, we were disqualified from backing freedom and democracy. "We have unclean hands," cried Congressman Vic Fazio, a California Democrat.[151]

The proposed liberal solution was always negotiation. Just as they believed in nuclear arms negotiations for their own sake, they believed in a "peace process" without regard to what its consequences might be. If a peace process were used by the Sandinistas to keep critics at bay while they consolidated their hold on power, this was ignored. And negotiation backed by implied force—as in Frederick the Great's dictum that "Diplomacy without arms is like music without instruments"—always seemed to strike liberals as needlessly warlike. And so they hailed one "peace plan" after another—the Wright Plan, the Contadora Process, the Arias plan—all the while providing only intermittent support for the Contras. It was impossible for any peace plan to fail in their eyes, since lack of progress was nearly always interpreted as evidence that new talks were now "urgent."

The Reagan administration did not spurn negotiations. Responding to congressional pressure as well as the institutional bias of the State Department, they engaged in one "process" after another, insisting only on good faith from the Sandinistas. But the Sandinistas' approach was more tactical. They were more than happy to emphasize negotiations. They lost nothing by talking, and it kept the Contras from receiving further aid.

In April 1985, just after Congress had voted down nonmilitary aid for the contras in a party-line vote, Nicaraguan president Daniel Ortega jetted off to Moscow—presumably to celebrate. Democrats were furious. "Politically-wise, he embarrassed us by his activities, to be truthful," Speaker O'Neill told the UPI wire service.[152] Six weeks later, a sufficient number of Democrats having been shamed into second thoughts, the aid package was approved.

Over the next two years, the Contras were able to inflict serious damage on the Sandinistas. As Humberto Ortega, Sandinista minister of defense, later acknowledged to historian Robert Kagan, "The war against the Contras was very, very hard."[153] And in the end, the pressure exerted by the Contra army, along with other inducements, forced the Sandinistas to schedule the elections that, of course, proved their undoing.

But the Reagan administration also struck a blow against itself by the dubious scheme to trade arms for hostages with Iran and funnel some of the profits to the Contras. Though top officials of the Reagan administration apparently believed sincerely that the scheme was not an arms for hostages swap, but rather an opportunity to thaw relations between Tehran and Washington, the Iranians certainly did not see it that way, and neither did the American people when the policy came to light. Sending the profits to the Contras certainly violated the spirit of the Boland Amendment, though it was never entirely clear—and for a variety of legal reasons was never ruled upon by a court—whether the diversion was actually illegal. Still, in the ensuing melee, the Contras were the losers, as the taint of the scandal was—justly or not—attached to them. The Democrats treated the matter as another Watergate complete with televised hearings and talk of impeachment. It weakened the president with the American people and thus with Congress.

Still, against all expectations, the Sandinistas were not preserved by this piece of luck. The Contras picked up new volunteers daily; the Nicaraguan economy was in shambles (30 percent unemployment,

33,700 percent inflation); and neighboring Central American nations were putting pressure on Managua to negotiate with the Contras and permit free elections. Moreover, as the decade drew to a close, the Soviet Empire began to fray, and the Sandinistas were forced to tack. They agreed to release some political prisoners and to hold elections on February 25, 1990, though they at first planned to rig them. Sandinista vice president Sergio Ramirez offered that it was "absurd" to believe that the Sandinistas would demilitarize and that "only those who are not in their right minds would think of requesting the surrender of the revolution."[154] But cheating became more and more difficult to manage as a flood of international observers deluged the small country to monitor the elections. Outsiders observed more than half the polling places.

As election day approached, most American liberals confidently predicted—or at least expected—a Sandinista victory. The *New York Times* intoned:

> Today's election in Nicaragua may finally break the fever. For 10 years, successive Republican Administrations in Washington have relentlessly distorted diplomacy and even traduced the law, all in fear of a small and impoverished neighbor in Central America....
>
> To look back is to recognize how badly America has bent itself out of shape over Nicaragua....The Contra war is over, the global Soviet threat has ebbed and the Sandinistas have lost much of their arrogance. If today's election is fair, and its results honored, Mr. Bush should relax and let Washington finally end its Nicaragua fixation.[155]

Those were not the words one would choose if an opposition victory were expected. NBC reporter Ed Rabel reported from Managua on February 21, "The election observers say the Bush Administration may have itself to blame for Daniel Ortega's rise in popularity among the voters. The reason, they say, is the U.S. military invasion in

Panama. That was a move that was widely denounced here in Nicaragua. It was a close race until the U.S. invaded."[156]

On *World News Tonight* on ABC, Peter Jennings told viewers, "For the Bush administration and the Reagan administration before it the [poll] hints at a simple truth: After years of trying to get rid of the Sandinistas, there is not much to show for their efforts."[157]

The *Nation* magazine declared that the Sandinistas "have no credible opposition." And Daniel Ortega himself proclaimed it "impossible" that he would be defeated. And ABC and NBC obviously agreed. Both predicted huge Sandinista victories.

How could they have been so wrong? Polls had predicted a comfortable Sandinista victory. And liberals, having learned nothing from seventy years of communism, simply accepted these polls at face value. They failed to consider that people might be too intimidated by the Sandinistas to tell the truth to pollsters. Further, they found it hard to believe that the Communists were not as popular in Nicaragua as they were in Washington, D.C., and New York City. And so they were blindsided. A *Nightline* program a day or so before the election was devoted entirely to the question: How would Washington cope with the Sandinistas once they had obtained a popular mandate?[158]

In the event, Violetta Chamorro, the principal opposition candidate, editor of *La Prensa* newspaper, won a resounding victory with 55 percent of the vote. (That was the official figure, though many observers believed the Sandinistas had actually fared even worse.)

The Contra war *was* over, but not, as the *New York Times* predicted, because Washington had finally surrendered and accepted the Communist Sandinista government, but because the Nicaraguan people had finally been given a chance to exercise their will.

Of the three nations that so preoccupied the United States during the 1980s—El Salvador, Grenada (briefly), and Nicaragua—all are now free and democratic. Had liberal policies prevailed, it is questionable whether any would be free today. And liberals would still be urging that the problem is poverty. (Now that Central America is no longer a Cold War hot spot, liberals seem to have forgotten about poverty there.)

The outcome in Latin America contradicted every favored liberal myth—that the Sandinistas were basically popular, that the guerrillas in El Salvador represented the "tide of history," that no progress toward peace was possible while the underlying causes went unaddressed—and yet liberals have soldiered on as oblivious to this rebuke as they were to the Vietnamese boat people and the Cambodian genocide. Not only has there been no reevaluation of the conflict in Central America, there has been an effort to punish those who were proved right.

When President George W. Bush nominated Otto Reich to be assistant secretary of state for Western hemisphere affairs, John Negroponte to be ambassador to the United Nations, and Elliott Abrams to serve on the National Security Council, the liberal hive began to buzz. The press began to repeat the word "controversial" about these nominees, and Senate Democrats predicted very tough confirmation fights (Abrams's post did not require Senate confirmation). All three had played leading roles in the Central America drama of the 1980s—Abrams as assistant secretary of state, Reich as director of the State Department's Office of Public Diplomacy, and Negroponte as U.S. ambassador to Honduras.

The *National Catholic Reporter* blasted all three nominees:

> Whatever qualities President Bush may recognize in his recent appointments to several key foreign policy posts, what much of the rest of the world sees is a brusque affront to the cause of human rights and a particular insult to several countries in Central America.
>
> At a time when El Salvador and Guatemala grope toward democracy while dealing with horrific memories of slaughter, torture, and genocide at the hands of official armies and paramilitary units, the United States is reward-

ing those who bore major responsibility for the U.S. role in the brutality.[159]

National Public Radio broadcast a lengthy story on Otto Reich alone. Noting that his father had fled first Hitler and later Castro, reporter Steve Inskeep labeled Reich a "fierce anticommunist" without also saying that he was a fierce anti-Nazi. (In the liberal world, being a fierce anti-Nazi is equivalent to be being a minimally decent human being—which is correct. But being a fierce anticommunist is seen as evidence of a character flaw.) NPR then quoted Reich circa 1985 on Nicaragua: "This is the first time in our history that we have been confronted with a hostile ideology attempting to establish itself on the continental Americas."[160] This quotation was supposed to demonstrate Reich was hot headed. For good measure, Inskeep mentioned that the Office of Public Diplomacy had been accused by "investigators" of conducting domestic propaganda. As it turned out, the "accusation" was never formal, and in any case it turned out to be groundless. Someone on Reich's staff had been accused of ghostwriting an op-ed for the *Wall Street Journal*. Investigation proved that the contraband article had never actually been written. But even if it had, the Office of Public Diplomacy was meant to build support for the president's policies. It's difficult to do that without persuasive op-eds.

Senator Christopher Dodd—the Connecticut Democrat nearly always wrong but never in doubt—told National Public Radio: "This is a person who I don't think meets the test of whether or not they're qualified to serve because of their past history, nor ideology, but how they managed offices they were in before."[161] During the 1980s, Reich had vigorously promoted the Reagan administration's Central America policies. That was what apparently disqualified him in Dodd's eyes.

Also seen as disqualifying was Reich's status as a Cuban refugee. This, they said, tainted Reich's views about Cuba. Has there ever been the suggestion, ever, that someone who fled Nazism or apartheid was morally suspect on that account? On the contrary.

Bill Goodfellow of the Center for International Policy condemned Reich's appointment as "a payoff to the Cuban-American community—the right-wing Cuban-American community Miami."[162] The Center for International Policy is a left-leaning think tank that features Castro-apologist Wayne Smith on its staff and Dessima Williams, identified on CIP's website as "former Grenadian ambassador to the Organization of American States," on its board of directors. Williams was the ambassador when Maurice Bishop's government controlled Grenada.

And so it went. Only the shock of September 11, 2001, derailed this absurd exercise in historical revisionism. It's been said that history is the polemic of the victor. Not, it seems, when liberal Democrats get hold of it.

# CHAPTER SEVEN

# Post-Communist Blues

*There is no error so monstrous that it fails to find defenders among the ablest of men.*  —LORD ACTON

WHEN LIBERALS ARE NOT CLAIMING (with straight faces) that they were "all Cold Warriors," they are proving that if the Soviet Union were to rise tomorrow like a Phoenix, they'd resume their former appeasement and America-bashing without skipping a beat.

In their response to two post–Cold War events—the Elian Gonzalez controversy and the terrorist attack of September 11, 2001—liberals demonstrated the same "useful idiot" syndrome that had characterized them throughout the second half of the Cold War. The tide of history has washed over them in a flood, yet they've stood their ground—scarcely noticing that they are all wet.

Before addressing those two revealing episodes, it's worth pausing to note the nostalgia for Communism that surfaced while the grave of Leninism was still freshly dug.

A number of reporters, particularly television correspondents, couched their dispatches from the newly free nations behind the old Iron Curtain in sour tones. ABC's Jerry King, for example, reported in October 1990, "East Germany is staggering toward unification, and may get there close to dead on arrival, the victim of an overdose of capitalism. Under communism, every worker was guaranteed a job. Under capitalism the goal is profit and companies like the old fashioned Brandenburg Steel Mill had too many employees to be cost effective."[1]

In addition to political bias, these kinds of reports betray an astonishing ignorance of economics. If the Brandenburg Steel Mill had too many employees to be "cost effective," then it was running at a loss. No firm that fails to earn a profit can provide high salaries (or any salaries) to its employees. The Communists compensated for this with subsidies. Since very few industries were profitable under communism, and since such a huge share of those profits were diverted to military uses, the meager subsidies were not sufficient to provide for thousands of workplaces—which is why East Germany was so poor relative to West Germany. But American reporters did not see this. Their hearts fluttered at the idea of a "guaranteed job for everyone."

"This is Marlboro country," announced Bert Quint on the April 11, 1990, CBS *Evening News*. "Southeastern Poland, a place where the transition from communism to capitalism is making people more miserable every day.... No lines at the shops now, but plenty at some of the first unemployment centers in a part of the world where socialism used to guarantee everybody a job."[2] A month later, Quint was driving home the same point:

> Communism is being swept away, but so too is the social safety net it provided.... Factories, previously kept alive only by edicts from Warsaw, are closing their doors, while institutions new to the East, soup kitchens and unemployment centers are opening theirs.... Here are the ones who

may profit from Poland's economic freedom: A few slick locals, but mostly Americans, Japanese, and other foreigners out to cash in on a new source of cheap labor.[3]

Connie Chung told CBS viewers in late 1991, "In formerly communist Bulgaria, the cost of freedom has been virtual economic disaster." She then tossed to Peter Van Sant, who explained that "Thousands of socialists rall[ied today] in Sofia, Bulgaria. It may look like a rally from Communism's glory years, but it's not. It's an expression of frustration, a longing for the bad old days when liberty was scarce but at least everyone had a job."[4] When were "Communism's glory years" one wonders? Is it possible to conceive of a reporter referring blandly to "Nazism's glory years"? Barbara Walters noted sadly that, "In the old Soviet Union, you never saw faces like these— the poor, the homeless, and the desperation of the Russian winter. Their numbers are growing. Tonight: Is this what democracy does? A look at the Russia you haven't seen before … the price of freedom can be painfully high."[5] Had Ms. Walters never heard of the millions slowly starved and the millions shot?

It seemed to gall many American reporters that the peoples of Eastern Europe were so desperate to be rid of a system that liberals considered to be, in many important respects, superior to the United States. Much of the reporting at the time had a grudging quality. CBS's Bob Simon, for example, seemed to scold the East Germans for their ingratitude in a 1990 dispatch: "Few tears will be shed over the demise of the East Germany army, but what about East Germany's eighty symphony orchestras, bound to lose some subsidies, or the whole East German system, which covered everyone in a security blanket from day care to health care, from housing to education? Some people are beginning to express, if ever so slightly, nostalgia for that Berlin Wall."[6] Perhaps, but most of them were American reporters.

ABC's Mike Lee was convinced that refugees fleeing Albania had been misinformed: "These refugees have been told little about the realities of life in the West, including the fact that some people sleep in the street.... They will soon learn that jobs are hard to find, consumer goods expensive, relatives in Albania will be missed. Many refugees, according to experts, will suffer from depression, and in some cases, drug abuse."[7]

Others would acknowledge that Communism was flawed, but only to highlight that capitalism is worse: "Under Communism," explained ABC's Jim Bitterman from Hungary, "neither unemployment nor the homeless existed officially, so help from the West was unnecessary. The charities were forced to go. But the Communists never really made the problems go away. So far, capitalism is just making them worse. Unemployment is expected to rise ten-fold in the next year. One out of every four people already live [sic] at the poverty line, and the government has no money to spare for social services."[8]

CNN's Steve Hurst reported from Moscow:

> Soviet people have become accustomed to security if nothing else. Life isn't good here, but people don't go hungry, homeless. A job has always been guaranteed. Now all socialist bets are off. A market economy looms, and the social contract that has held Soviet society together for seventy-two years no longer applies. The people seem baffled, disappointed, let down. Many don't like the prospect of their nation becoming just another capitalist machine.[9]

If freedom meant economic woe it also brought social regression, at least through the eyes of the *Los Angeles Times*. "Ten months after the new Germany emerged," wrote Tamara Jones in 1991, "women in the eastern sector are coming to the stunning realization that, in many ways, democracy has set them back forty years."[10] *U.S. News and World Report* concurred: "Like many other women in what used to be the German Democratic Republic, she worries that political liberalization has cost her social and economic freedom.... The kindergartens that cared for their children are becoming too expensive, and

West Germany's more restrictive abortion laws threaten to deny many Eastern women a popular form of birth control.... East Germany's child care system helped the state indoctrinate its young, but also assured women in the East the freedom to pursue a career while raising a family."[11]

Robin Wright of the *Los Angeles Times* added an academic-sounding gloss to this same idea in an op-ed for the *Philadelphia Inquirer*:

> Open societies, it turns out, haven't been as generous as socialism and communism to women who want to serve in public office. From Albania to Yemen, the number of women in power plummeted after the transition from socialist governments, which sought to develop female as well as male proletariats. As those governments died, so went the socialist ideals of equality and the subsidies for social programs that aided women. In many countries, traditional patriarchal cultures resurfaced.[12]

And in 1994, ABC News was still attributing the miseries of post-communist societies to capitalism. Richard Gizbert reported:

> In Bucharest, the train station provides not only a means of transport, but a place of refuge from the cold. In the shadow of kiosks selling American cigarettes are the casualties of Romanian capitalism... the poor, the homeless, the children.... Most of Bucharest's 500 homeless children have parents, but the days of state subsidies for large families are gone. Poor families... are falling apart. Children are abandoned. Others run away.[13]

You must harbor a deep animus toward the United States to frame the story of Bucharest's homeless children with the words "In the shadow of kiosks selling American cigarettes...." The cigarettes are irrelevant to the story, except to imply some sort of American malfeasance or nonfeasance.

Former NBC anchor John Chancellor had bad news to deliver about Russia, but he was careful to ensure that his listeners didn't blame the wrong culprit: "It's short of soap, so there are lice in hospitals. It's short of pantyhose, so women's legs go bare. It's short of snowsuits, so babies stay home in winter.... The problem isn't communism; nobody even talked about communism this week. The problem is shortages."[14] At other times, the message was implicit. On NBC's *Nightly News*, Bob Abernethy reported, "[The Soviet] Congress changed the Soviet constitution to permit limited private ownership of small factories, although laws remain against exploitation of everyone else."[15] Think about that.

## THE DOMESTIC COLD WAR NEVER ENDED

The Elian Gonzalez affair provided ample evidence that American liberals had learned nothing about the nature of communism from history; nothing from the heart breaking accounts of refugees; and finally, nothing from its spectacular implosion in the Eastern bloc.

In November 1999, on Thanksgiving Day, fishermen off the Florida coast found a five-year-old boy floating in an inner tube. He was one of three survivors of a boat full of refugees from Castro's Cuba. The boat had sunk in rough waters, drowning both Elian Gonzalez's mother and her boyfriend. Miami relatives who had been expecting him immediately took in the child.

Over the course of the next several months, America's domestic Cold War divisions flared anew. Liberals demonstrated the same ignorance and credulity, and yes, cordiality toward Cuba that they had shown so regularly toward all Communist regimes from 1967 to 1991. And the Clinton administration proved abundantly willing to permit Castro to control events.

The boy was cared for by his great-uncle Lazaro Gonzalez and his young cousin Marisleysis. Elian's parents had been divorced. His father, Juan Miguel, had remarried and was living in Cardenas, Cuba. The Miami relatives were expecting Elian (and his mother) because

Elian's paternal grandfather (who lived in the home of Elian's father) had phoned two days before Elian washed ashore to alert them to expect the pair. Telephone records confirmed the call. Is it conceivable that Elian's grandfather knew of Elian's mother's plan but the boy's father did not? Perhaps, but it's extremely unlikely.

Castro (whose own six-year-old son had been spirited to America many years before) was incensed by Elian's safe arrival in Miami and launched a campaign of agitation to have him repatriated. Hundreds of thousands of Cubans were forced to march in "spontaneous" demonstrations for Elian's return to Cuba (schools and factories were closed for these events). Elian's desk at school became a national shrine, and his picture was plastered on billboards all over the country. Cuban television featured talk shows every night on one subject: Elian. (Elizabet Broton, the child's mother, who died attempting to free herself and Elian, was, of course, denounced as a traitor.)

From the moment Elian's existence came to public attention, liberals in the press and government betrayed a crude contempt toward Cuban-Americans and a strong bias in favor of sending the boy back to Cuba and his father. They demonstrated no understanding that totalitarian regimes regard it as a crime to flee, commonly punish the relatives of those who defect, and generally apply strong-arm tactics to ensure that people "tow the party line."

And so the Clinton administration sent a couple of immigration officers to Havana to question Elian's father, Juan Miguel Gonzalez, about his wishes for the child. What in the world did the Clintonites think the man would say under the circumstances? With Castro making the boy the biggest issue on the island since Soviet nuclear missiles were stationed there, was Juan Miguel truly at liberty to say, "Gee, since he's made it to the U.S. and Uncle Lazaro is eager to raise him, I'd like him to stay"? Other survivors of Elian's capsized craft said that Juan Miguel and his new wife and baby were hoping to come to the U.S., too. Juan Miguel certainly could not have been frank about that. It's hard to know whether these accounts are accurate, but they are consistent with the efforts of so many thousands of Cubans to flee.

Juan Miguel Gonzalez, wrote the *Washington Post*'s Richard Cohen, had acted like "a typical father." The very notion that Juan Miguel's actions could be evaluated without reference to the despotism under which he lived was absurd. But besides, Mr. Gonzalez's actions were anything but typical. He waited four months before coming to the United States to recover his son. Once on American soil, he traveled to a different city from the one where Elian was living. This is not to imply that he was unfeeling or indifferent—merely that he was, without doubt, under orders.

The Reverend Joan Brown Campbell, a leader of the National Council of Churches, was all over television during the Gonzalez drama, describing the great love between the father and his son and denying point blank that Juan Miguel was under any pressure from Castro. Almost no one ever bothered to report *her* history, which showed a great love for Castro and communism. Though she headed a nominally Christian organization, she once offered the view that "If you look at the Nazi regime, you see in it the philosophy of Christian superiority."[16] On another occasion, she appeared with Castro at a Havana rally telling the crowd, "We ask you to forgive the suffering that has come to you by the actions of the United States....It is on behalf of Jesus the liberator that we work against this embargo."[17]

Just as they misconceived the pressures on Gonzalez to say only what Castro approved, some reporters expressed puzzlement at the actions of Elizabet Broton, the boy's mother. "Why did she do it?" asked ABC's Jim Avila. "What was she escaping? By all accounts this quiet, serious young woman who loved to dance the salsa, was living the good life, as good as it gets for a citizen of Cuba." Avila could only conclude that Broton had made a terrible choice: "An extended family destroyed by a mother's decision to start a new life in a new country, a decision that now leaves a little boy estranged from his father and forever separated from her."[18] Other reporters circulated gossip suggesting that Broton was unstable and a party girl.

As for the Cuban-American community in Miami, it was clear that they were extremists and fanatics. "Communism Still Looms as

Evil to Miami Cubans," explained an immortal *New York Times* headline.[19] NBC's Katie Couric offered this waspish comment: "Some suggested over the weekend that it's wrong to expect Elian Gonzalez to live in a place that tolerates no dissent or freedom of political expression. They were talking about Miami...."[20]

Those who opposed returning Elian to Cuba were the ones tagged as authoritarian. John Quinones of ABC reported, "It seems like such a contradiction that Cubans, who profess a love of family and respect for the bond between father and son, would be so willing to separate Elian from his father. But in Miami it's impossible to overestimate how everything is colored by a hatred of communism and Fidel Castro. It's a community with very little tolerance for those who disagree."[21] It is also the only minority community in America that receives brickbats from the press. These Spanish-speaking immigrants are entitled to none of the special consideration that attaches to other immigrant and minority populations because the Cubans vote Republican and are strongly anticommunist.

"Hating" communism should be the minimum expected of any civilized human being. Instead, the liberal press regarded it as some sort of pathology unique to South Florida. And Quinones got more than that wrong. The Cuban-Americans of Miami never agitated to separate Elian from his father. Their preferred solution was to have the whole Gonzalez family, including Juan Miguel's new wife and child and Elian's grandmothers, come to the U.S. to discuss matters. Only then could the father's true desires be ascertained. Sure, if any or all had elected to defect, certainly the Miami community would have cheered. But their principal aim was to prevent Castro from stage-managing the entire affair through intimidation. But Juan Miguel's true wishes were never discovered, because the people in charge of U.S. policy at the time, President Clinton and Attorney General Janet Reno, were every bit as soft on Castroism as the press.

If Bill Clinton had publicly invited Juan Miguel Gonzalez and his family to the U.S. to settle the matter, he would have satisfied the Miami Cubans and shifted the focus of debate back where it belonged—on the repressive Cuban government. If Castro had agreed to permit the trip, and it turned out that Gonzalez was one of the rare true believers in communism on the island—so be it. At least the question of strong-arming would be set aside. If, on the other hand, Castro had refused permission for the family to travel to America, he would have been proving what so many were at pains to point out—that Castro's biggest fear is always defection. This was the course recommended by then presidential aspirant George W. Bush (which signaled an understanding of the nature of totalitarianism). But Bill Clinton, while claiming throughout the protracted controversy that he was considering only the best interests of the child, in fact put the Secret Service, the Immigration Service, the Justice Department, and the courts, in the service of Fidel Castro. With this government's assistance, Juan Miguel was kept under constant watch by Cuban agents when he traveled in the United States. He was never free to so much as stroll down Washington D.C.'s Connecticut Avenue unaccompanied.

Ms. Reno, whose previous gesture on behalf of children led to their incineration at Waco in 1993, simply decreed that the boy belonged with his father. At one point in the drama, the National Council of Churches, which worked hand in glove with Castro throughout the Elian battle, chartered a plane to fly Elian's two grandmothers to Miami. (Later, when Juan Miguel finally came to the U.S., also using the NCC chauffeur service, the grandmothers were placed under house arrest in Cuba. This was virtually ignored by the press.)

Elian's grandmothers met with the boy at a neutral site, the home of longtime Janet Reno friend Sister Jeanne O'Laughlin. Much was made of this meeting until O'Laughlin announced that she had changed her view. With the exception of Cokie Roberts of ABC and several conservative columnists, few media talking heads paid any attention. And certainly Janet Reno paid no attention. O'Laughlin witnessed the "trembling, furtive looks" the grandmothers exchanged, their "ice cold hands," and the overall "atmosphere of fear" that characterized their visit. They, too, were accompanied at all times by government agents. It made an impression on O'Laughlin—but the press buried it.[22]

Castro's boosters in the U.S. Congress took notice though. California Democrat Maxine Waters was biting: "I am bewildered. Never in my wildest imagination would I think that a nun who was supposed to be a neutral party would undermine that neutrality."[23] It was the Miami Cubans who had "little tolerance for those who disagree"?

On the CBS *Evening News*, Byron Pitts reported, "Six weeks ago this community embraced a boy who had watched his mother die at sea. Tonight there is fear that the embrace has become a choke hold." On the CBS *Early Show*, Bryant Gumbel, never one for subtlety, asked an interviewee, "Cuban-Americans...have been quick to point fingers at Castro for exploiting the little boy. Are their actions any less reprehensible?"[24]

Over at *Newsweek*, Evan Thomas and Joseph Contreras were sure that "with the right nurturing, Elian Gonzalez may overcome his nightmares, but he has been scarred and prematurely aged, first by losing his mother in a terrifying accident at sea, then by the grotesque spectacle of his martyrdom in Miami."[25]

*Time* magazine's Tim Padgett mistook the media's interpretation of events for America's: "The 'banana republic' label sticking to Miami in the final throes of the Elian Gonzalez crisis is a source of snide humor for most Americans. But many younger Cuban-Americans are getting tired of the hard-line anti-Castro operatives who have helped manufacture that stereotype...but [are] ever ready to jump on expensive speedboats to reclaim huge family estates the moment the old communist dictator stops breathing."[26] Padgett practically labeled them as "counterrevolutionaries"—the worst insult in the communist vocabulary—by writing that "the older hard-liners, despite their protestations of U.S. patriotism, are still steeped in the authoritarian political culture that existed in Cuba long before Castro took power in 1959."[27]

On what grounds did so many commentators conclude all of this? Because the Miami Cuban community was nearly unanimous in opposing Elian's return to Cuba before his asylum petition could be adjudicated?

Many liberals, including the president of the United States, simply could not understand what all the fuss was about. At a press conference President Clinton allowed that Elian would have more "economic opportunity" here, but he apparently could think of no other reason that the child might be better off in the United States.[28]

Peter Jennings thought the matter fraught with uncertainty. "Beyond the questions of custody, the Cuban-American community in Miami has always argued, almost every day in fact, that Elian Gonzalez would have a better life here in the United States than in Cuba. It's been argued before, and there's not a simple answer."[29] There isn't?

At least Clinton seemed to think there was *some* advantage to life in the United States over Communist Cuba. *Newsweek*'s Eleanor Clift would not even go that far. Speaking on *The McLaughlin Group*, Clift said, "To be a poor child in Cuba may in many instances be better than being a poor child in Miami and I'm not going to condemn their lifestyle so gratuitously."[30] It must come as news to Cubans suffering under their tyrannical government that they've made a "lifestyle" choice. But perhaps Clift spoke in haste and was misinterpreted? Offered an opportunity to flesh out her views on *The O'Reilly Factor*, she offered pure agitprop: "I can understand why a rational, loving father can believe that his child will be protected in a state where he doesn't have to worry about going to school and being shot at, where drugs are not a big problem, where he has access to free medical care and where the literacy rate is I believe higher than this country's."[31]

Others were quick to cite America's crime problem as one of many reasons that Elian might be better in Cuba. Larry King asked Tipper Gore: "One of the things that Elian Gonzalez's father said that I guess would be hard to argue with, that his boy's safer in a school in Havana than in a school in Miami. Good point?"[32] Brook Larmer and John

Leland of *Newsweek* wrote: "Elian might expect a nurturing life in Cuba, sheltered from the crime and social breakdown that would be part of his upbringing in Miami." In Cuba, they summarized, "The boy will nestle again in a more peaceable society that treasures its children."

Drivel. Cuba indoctrinates its children—thoroughly. And Cuba is not more "peaceable" than the United States. It's just that all of the violence comes from the state. Cuba is a thoroughgoing totalitarian nightmare, not some sort of uptight Bermuda. Americans, crime ridden or not, are not climbing into boats attempting to get to the "peaceable society that treasures its children." It is a crime to attempt to flee Cuba, to start a labor union, or to form a political party. Trials are conducted in secret—and in political cases their outcomes are a foregone conclusion. Phones are tapped and children are encouraged to rat out their parents for insufficient communist consciousness. Parents can actually lose custody of their kids if a neighborhood committee determines that they are not adequately instilling a "communist personality" in the child. Teachers are required to keep records not just of each child's academic progress but of his political progress. Any deviation from party orthodoxy is noted.

Many reporters were probably ignorant of this aspect of Cuban childrearing. But even among those familiar with it, it caused no ruffled feathers. *Washington Post* reporter John Ward Anderson wrote: "Starting in the first grade, all Cuban children are members of the Young Pioneers—a group that Cuban exiles claim imparts communist ideology, but which parents say also teaches social skills and responsibility. Although they begin each day reciting 'Pioneers for communism will be like Che!' few children give it much thought, parents said." (These are the same liberals who would faint if a Psalm were read in an American public school.)

CBS's Dan Rather actually offered an on-air character reference for Fidel Castro: "While Castro, and certainly justified on his record, is widely criticized for a lot of things, there is no question that Castro feels a very deep and abiding connection to those Cubans who are still in Cuba. And, I recognize this might be controversial, but there's little doubt in my mind that Fidel Castro was sincere when he said, 'Listen, we really want this child back here.'" The Maximum Leader Truly Loves His People. If he loses his day job, Rather could compose Cuban billboards.

Elian was served up to Castro—and in the crudest possible way. The attorney general and the president ordered a predawn armed raid of the Gonzalezes' Miami home and tore Elian, screaming with fear, from the house at gunpoint. The Gonzalez family had committed no crime and was in violation of no court order. The Clinton administration apparently never considered simply knocking on the door and asking for the boy. Brandishing weapons, kicking and stomping an NBC film crew, threatening to shoot if anyone moved—it is difficult to imagine any other scenario in which American government agents could behave this way toward American citizens and get such a free pass from the press.[33] The tactics caused barely a ripple because so many in the press openly approved of Reno's action—though much of the nation was outraged. The still photos of the raid caused plenty of trouble for the Clinton administration and did grace the covers of the major newsweeklies. Video would have been far worse. But the NBC cameraman on the scene wound up in the hospital after his encounter with Reno's shock troops. Though two civil libertarians of liberal bent, Alan Dershowitz and Lawrence Tribe, denounced the raid as illegal and unconstitutional, their critiques received little attention from a press that was satisfied with the result.

The *New York Daily News* cheered that Elian had been removed from "the Miami mob scene."[34] Rick Bragg, Miami bureau chief for the *New York Times,* confessed, "Some people think hell is a place where you wake up in the morning in a bed of coals. I think it's where you wake up and find out you'll be writing about Elian for the next 643 days."[35] The *St. Petersburg Times* did not mind the predawn raid because Elian "was manipulated and brainwashed by his Miami relatives . . . [they had] abused this child long enough."[36]

*New York Times* columnist Thomas Friedman wrote: "Yup, I gotta confess, that now-famous picture of a U.S. marshal in Miami pointing an automatic weapon toward Donato Dalrymple and ordering him in the name of the U.S. government to turn over Elian Gonzalez warmed my heart."[37] Speaking on *Inside Washington*, *Newsweek*'s Evan Thomas thought Reno had a heart of gold: "I think Reno really comes through this as somebody who may have made mistakes, but was principled about it, and unlike most people in Washington, who are trying to figure out the political aspect of it, seemed quite apolitical about it."

Some time in the not too distant future, Fidel Castro will die, and without him the Communists will most probably be toppled. When that happens, the bitter cruelty that Communism has imposed on the Cuban people will at last come to light. And the attitudes of American liberals toward that regime will be revealed for what they are—a disgrace.

## THE SEPTEMBER 11 AFTERMATH: STILL BLAMING AMERICA FIRST

Just three days after the hijacked planes exploded into the World Trade Center and the Pentagon, leftist demonstrators took to the streets of Washington, D.C., carrying placards and handouts saying, "No Eye for an Eye," "No More War," and "No Further U.S. Violence."[38] Even when the financial district of New York had taken a direct hit with thousands murdered, the leftist impulse was to condemn "U.S. violence."

That this inclination was still alive after the carnage of September 11 is testimony to its durability. Americans should think twice before believing that "everything has changed" since that day.

Those anonymous leftists were first out of the box but they were hardly alone. Katha Pollitt demonstrated the reliable theme of America-loathing that informs much leftist thinking. Her column in the *Nation* magazine after the terror attacks expressed her repugnance at the American flag:

> My daughter, who goes to Stuyvesant High School only blocks from the World Trade Center, thinks we should fly an American flag out our window. Definitely not, I say: The flag stands for jingoism and vengeance and war.... It seems impossible to explain to a 13-year-old, for whom the war in Vietnam might as well be the War of Jenkins's Ear, the connection between waving the flag and bombing ordinary people half a world away back to the proverbial stone age. I tell her she can buy a flag with her own money and fly it out her bedroom window....
>
> I've never been one to blame the United States for every bad thing that happens in the Third World, [here comes the "but"], but it is a fact that our government supported militant Islamic fundamentalism in Afghanistan after the Soviet invasion in 1979....
>
> Bombing Afghanistan to "fight terrorism" is to punish not the Taliban but the victims of the Taliban, the people we should be supporting. At the same time, war would reinforce the worst elements in our own society—the flag-wavers, and bigots and militarists.[39]

Sometimes, delivering people from oppressive governments is the very best way to "support" them. Clearly, Ms. Pollitt felt more threatened by the "worst elements" in our society, including "flag-wavers," than by the terrorists.

Novelist Barbara Kingsolver, writing in the *San Francisco Chronicle*, teed off Jerry Falwell's unfortunate comments blaming the events of September 11 on homosexuals and abortion supporters to launch a full scale rant: "In other words, the American flag stands for intimidation, censorship, violence, bigotry, sexism, homophobia, and shoving the Constitution through a paper shredder. Who are we calling terrorists here?"[40] Falwell and Pat Robertson both made hotheaded comments about God's retribution for America's sinfulness on the day of the attack, but Falwell later amended his remarks and apologized.

Susan Sontag, who had sidled up to Ho Chi Minh during the Vietnam War, took aim at her own country again after September 11. Writing in the *New Yorker*, she scorned the "unanimity of the sanctimonious, reality-concealing rhetoric spouted by American officials and media commentators." Like Pollitt, Sontag would not be caught dead flying a flag from her window—well, not an American flag anyway. (Have the antiwar protesters from the Sixties held on to their VC flags?) "We are told again and again . . . [that] our country is strong," she commented acidly. "I, for one, don't find this entirely consoling."[41]

Professor Edward Said of Columbia University, sometime member of the Palestine National Council, offered that Arab hatred of the U.S. was quite unsurprising in light of American "support for the thirty-four-year-old Israeli occupation of Palestinian territories."[42] (Why didn't Osama bin Laden's men fly planes into Kuwait City then? Kuwait expelled 500,000 Palestinians after the Gulf War.) *Salon.com*'s Gary Kayima agreed: "As long as millions of Islamic and Arab people hate America because of its Mideast policies, we will be in danger."[43] Should the U.S. adopt policies agreeable to the worst elements of the Islamic world in order to avoid danger? Wouldn't the words "cowardly" and "dishonorable" fit us if we did?

Essayist Vivian Gornick wrote in the *Village Voice*, "A military strike? Where? What? When? Above all, against whom? If you hit them in Iraq, they'll regroup in Libya. If you squash them in Libya, they'll rise up in Afghanistan. They have struck us, and in their strike announced: We'd rather die—and take you with us—than go on living in the world you have forced us to occupy. Force will get us nowhere. It is reparations that are owing, not retribution."[44]

The Episcopal Church urged its members to "wage reconciliation," because "The affluence of nations such as our own stands in stark contrast to other parts of the world wracked by crushing poverty which causes the deaths of 6,000 children in the course of a morning."[45] Just as with every other conflict the United States has been engaged in over the past forty years, many liberals stood ready to cite poverty as a justification for any outrage—even terrorism. Not all liberals allowed their knees to jerk at the invocation of poverty though. Sean Wilentz, Princeton professor and Clinton defender, observed, "To say that poverty explains terror is to slander those caught in poverty who choose to lead worthy lives. [Terrorists] are not the oppressed, but they are parasites on oppression."[46] Wilentz also noted that based on the biographies of the September 11 attackers, the logical inference would be that "money, education, and privilege" cause terrorism.[47]

Joel Rogers of the *Nation* condemned the September 11 attacks, but felt constrained to point out that "our own government, through much of the past fifty years, has been the world's leading 'rogue state.' The U.S. has taken the lives of literally hundreds of thousands, if not millions, of innocents, most of them children...merely listing the plainly illegal or unauthorized uses of force the U.S. was responsible for...would literally take volumes."[48]

Noam Chomsky was ready to fill volumes. Though (or is it because?) Chomsky is a white-hot America basher who is quite popular on the college lecture circuit. Though Chomsky rarely appears on television, his complex conspiracy theories and twisted history reach a great many Americans. He published a short book called *9-11* in question and answer format to air his views about the war on terrorism. A sample:

> *Your comment that the U.S. is a "leading terrorist state"*
> *might stun many Americans. Could you elaborate on that?*
>
> The most obvious example, though far from the most
> extreme case, is Nicaragua. It is the most obvious because
> it is uncontroversial, at least to people who have even the
> faintest concern for international law...the U.S. is the only

country that was condemned for international terrorism by the World Court.... I don't know what name you give to the policies that are a leading factor in the death of maybe a million civilians in Iraq and maybe half a million children, which is the price the Secretary of State says we're willing to pay. Is there a name for that? Supporting Israeli atrocities is another one.[49]

Michael Lerner, editor of *Tikkun* magazine (and reputedly one of Hillary Clinton's favorite spiritual thinkers), wrote, "We need to ask ourselves what is it in the way that we are living, organizing our societies, and treating each other that makes violence seem so plausible to so many people? And why is it that our immediate response to violence is to use violence ourselves—thus reinforcing the cycle of violence in the world?"[50] Lerner is not really asking us to think about these matters. He's already done all the asking and found the following answer:

> We may tell ourselves that the current violence has "nothing to do" with the way that we've learned to close our ears when told that one out of every three people on this planet does not have enough food.... We may tell ourselves that the suffering of refugees and the oppressed have nothing to do with us.... But we live in one world, increasingly interconnected with everyone, and the forces that lead people to feel outrage, anger, and desperation eventually impact on our own daily lives. The same inability to feel the pain of others is the pathology that shapes the minds of these terrorists.[51]

This is the sort of vapid blather that often travels under the name "liberalism." It contributes nothing to our understanding of the terror threat, nor to what motivates Islamic extremists to hate us, nor to what can be done about it. Lerner takes it as an article of faith that all suffering everywhere is somehow America's fault—even if only because we don't feel anguished enough about it. But taking it one step further, Americans actually do care a great deal about world poverty. The level of American engagement in the Third World is quite high. America offers foreign aid to the poorest nations on earth (including repeatedly forgiving our loans to them), and sponsors or largely funds the Peace Corps, the Agency for International Development, the World Bank, the International Monetary Fund, and the work of hundreds of nongovernmental private agencies. We have also gone to war several times in recent memory to prevent people in the Third World from starving (Somalia), or to prevent them from being ethnically cleansed (Bosnia and Kosovo). Reading Lerner probably subtracts from the sum of world knowledge.

Leon Fuerth, who had served as Vice President Al Gore's national security adviser, worried publicly on September 16 whether the Bush administration might take steps without consulting with the United Nations. "One thing is clear. We will need the support of others. To gain that will we be more likely than in the recent past to look for collective solutions rather than to go our own way when it suits us?"[52] It's not clear what Fuerth considers the recent past, but certainly the Clinton administration scarcely buttered its toast in the morning without notifying the General Assembly, and the George H. W. Bush administration was widely praised for assembling a large, international coalition to expel Iraq from Kuwait. What is not clear at all is whether international cooperation is the best guarantor of American security.

David Westin, president of ABC News, cautioned his on-air talent against wearing American flag lapel pins as this would constitute "taking sides." During a question and answer session with journalism students at Columbia, Westin was asked whether the Pentagon was a legitimate target. He gave the following answer:

> I actually don't have an opinion on that, and it's important that I not have an opinion on that as I sit here in my capacity right now. . . . I can say the Pentagon got hit. I can say

"this is what their position is," "this is what our position is," but for me to take a position this was right or wrong, I mean, that's perhaps for me in my private life. Perhaps it's for me dealing with my loved ones. Perhaps it's for my minister at church. But as a journalist I feel strongly that's something that I should not be taking a position.[53]

This fatuous neutrality doesn't seem to inhibit ABC in other contexts. But in any case it's a fraud. It isn't objectivity or neutrality that journalists like Westin guard so jealously. They do not hesitate to condemn their country when they think censure is justified (namely constantly). No, what shames them among their peers is to be caught sympathizing with their country, or indulging patriotic feelings. Still, those feelings did surface after September 11—if only briefly. (After all, if the French were momentarily pro-American, it isn't altogether surprising that liberals were as well.)

The New York Philharmonic Orchestra, for example, not usually known for its patriotic feeling, rummaged in the basement to find the American flag that had flown on its stage throughout World War II and displayed it proudly. Before performing its first post-terror concert, the orchestra played the "Star Spangled Banner"—with all of the musicians standing.

A number of leading popular entertainers—including Mariah Carey, Tom Cruise, Tom Hanks, Julia Roberts, and Bruce Springsteen—raised money for the families of the victims of September 11 in a telethon called "America: A Tribute to Heroes." On the other hand, they declined to commit themselves to USO tours to entertain American troops abroad. "Most rock musicians said that since they did not know how current events were going to play out, it was hard for them to tell whether they would be supporting their country's actions or criticizing them."[54] The shadow of Vietnam still darkens the landscape.

From the rabid end of the political spectrum came cartoonist Ted Rall, who dipped his pen in bile to write:

We've been treated to some astonishingly vile images over the last two weeks: office workers hurling themselves into a hundred-floor-high abyss. A gaping, smoldering hole in the financial center of our greatest city. George W. Bush passing himself off as a patriot, even as he disassembles the Constitution with the voracious glee of a piranha skeletonizing a cow.

Now we know why 7,000 people [initial estimates turned out to be high] sacrificed their lives—so that we'd all forget how Bush stole a presidential election.... Bush has capitalized on a nation's grief, confusion and anger to extort a political blank check payable in young American blood.[55]

Michael Moore, sometime filmmaker and all-purpose leftist, seemed troubled by the terrorists' choice of cities: "Many families have been devastated tonight," he opined on September 12. "This just is not right. They did not deserve to die. If someone did this to get back at Bush, then they did so by killing thousands of people who did not vote for him! Boston, New York, D.C.... these were places that voted against Bush!"[56] This is grotesque by any standard. Later, Moore decided that the attacks were deserved. "We have orphaned so many children...with our taxpayer funded terrorism" that "we shouldn't be too surprised when those orphans grow up and are a little wacked in the head."[57]

At a memorial service for victims of the September 11 attacks held in San Francisco on September 17, a former city supervisor, Amos Brown, intoned:

America, America. What did you do—either intentionally or unintentionally—in the world order, in Central America, in Africa where bombs are still blasting? America, what did you do in the global warming conference when you did not embrace the smaller nations? America, what

did you do two weeks ago when I stood at the world conference on racism, when you wouldn't show up? Oh America, what did you do?[58]

Norman Mailer, true to form, advised that "Americans should
reflect on and try to understand why so many people feel a revulsion
toward the U.S." Mailer added that in much of the world, the U.S. is
seen as the agent of "cultural and aesthetic repression." That is doubtful for the world in general, but particularly for the Islamic world.
There the United States is seen as a cultural threat; a liberator (for good
or ill), not a repressor. And that is one reason the United States is so
hated. It represents modernity, pluralism, sometimes, and libertinism—
while repression is what Middle East societies are all about.

Robert Scheer of the *Los Angeles Times* (a one-time admirer of
North Korea) wrote a column "congratulating" Bill Maher (the talk
show host who had drawn criticism for his post–September 11 comments); Congresswoman Barbara Lee, the sole member of Congress
to vote against giving President Bush authority to wage the war on
terror; Susan Sontag (presumably for everything); and Richard Gere
(who made insipid comments about offering the terrorists love and
understanding). He praised them for their independence and not
"blindly accept[ing] the actions taken in our name by the government."[59] But the richest part of Scheer's piece was this: "To understand the limits of government-sponsored 'unity,' we might ask the
soldiers of the old Soviet Union. They marched with their pledges and
anthems into . . . Afghanistan two decades ago, while at home the dissent that could have saved them from military and economic disaster
was systematically squelched."[60] The comparison is ridiculous. There
is no "squelching" of dissent in the United States. Certain people who
are not used to it were merely criticized. That is evidence of free
speech, not its absence. Further, for Scheer to compare the U.S. action
in Afghanistan to the Soviet invasion of that country (which, lest we
forget, loosed so many of the evils that now confront us in the form

of Islamic terrorism) is really quite sinister. It is also brazen coming from a man who almost never found anything critical to say about the Soviet Union while it existed.

The *Chronicle of Higher Education* ran a symposium on the terror attacks, which (along with much sensible commentary) included this offering from Professor David P. Barash of the University of Washington:

> If it is human nature to seek revenge, then it seems that an equally human nature motivated the perpetrators, who perceive themselves to be seeking revenge. If the United States, in its righteous anger, will "make no distinction between terrorists and those who harbor them"—in the words of President Bush—then, in view of the fact that many people consider the United States to be a terrorist state, weren't the perpetrators following just such a policy in attacking innocent civilians—making no distinction between their view of the terrorists (our government, our country) and those who harbor them (ourselves)?[61]

Even aside from the sentiments expressed, it's remarkable to reflect that people who write like that are teaching the young.

At Harvard, most of the student body probably backed the president, but some carried signs saying, "War is also terrorism." At Brown University, the faculty circulated a "curriculum guide" on handling the issue in class that recommended "understanding why people resent the United States."[62] The fire department in Berkeley, California, was forbidden to fly the American flag for fear of inciting the America-hating denizens of that city. A chorus of disapproval forced a reversal of policy. Poet Robin Morgan circulated an e-mail that described the attacks as stemming from "a complex set of circumstances including despair over not being heard."[63] Professor Eric Foner of Columbia University, historian and lifelong leftist, wasn't sure "which is more frightening: the horror that engulfed New York

City or the apocalyptic rhetoric emanating daily from the White House."[64]

Alison Hornstein, a student at Yale, penned a "My Turn" column in *Newsweek* that limned the moral void at the center of her ultraliberal education. Her Yale classmates, she found, could think of many reasons for the terrorists' hatred of the U.S., but could muster very little anger or indignation. "My generation is uncomfortable assessing, or even asking, whether a moral wrong has taken place.... The explanations students and professors give for the September 11 attacks—extreme poverty in the Middle East, America's foreign policy [i.e. support for Israel] in that region and religious motivation—are insightful, but they cannot provide absolution for wrongdoing."[65]

Bravo to Ms. Hornstein. But how disturbing that she stands out among Yale students.

History professor Richard A. Berthold told his University of New Mexico class on September 11, "Anyone who can blow up the Pentagon has my vote."[66] (Berthold later apologized, saying, "I was simply being at the moment an incredibly insensitive and unfeeling jerk."[67] He is too kind.)

On the other hand, Kenneth Hearlson, professor of political science at Orange Coast College in California, stood before his class and demanded "I want to see the Arab world stand up and say, 'This is wrong.'" He was suspended.[68]

Duke University shut down a professor's website because he posted an article that urged a strenuous military response to terrorism.[69] The administrators at Pennsylvania State University agreed, warning one of their professors that recommending a military response was "insensitive and perhaps even intimidating."[70] (What are military strikes supposed to be?) At Central Michigan University, students were admonished for adorning their rooms with posters of eagles and American flags.[71] At San Diego State University, an Ethiopian student who knew Arabic overheard several Arab students reveling in the success of Osama bin Laden's attack on America.

"They were happy," he reported. The Ethiopian student challenged them, saying "You should be ashamed." For this he was put "on warning" by the university administration.[72]

Former president Bill Clinton, who had traveled to Africa and other destinations during his presidency apologizing for America's past sins, kept to that theme after September 11. Addressing an audience at Georgetown University, Clinton offered a tour d'horizon of American and Western guilt, starting with the Crusades and ending— well, never ending. "Here in the United States," he explained, "we were founded as a nation that practiced slavery, and slaves quite frequently were killed even though they were innocent. This country once looked the other way when a significant number of Native Americans were dispossessed and killed to get their land and their mineral rights or because they were thought of as less than fully human. And we are still paying the price today."[73]

"And we are still paying the price today." Every word Clinton spoke (and this *is* noteworthy) was true except for the last sentence, which makes his entire point twaddle. History, including American history, is replete with injustice, slavery, and cruelty. But to suggest, as Clinton clearly did, that the Islamic fanatics who attacked us were motivated in any way by a sense of outrage about American slavery or ill treatment of the Indians is risible. Fifteen of the nineteen terrorists of September 11 hailed from a nation, Saudi Arabia, that outlawed slavery only in 1962 (and has arguably practiced it informally since then). Or if Clinton's idea is that America's past sins make us fair game for anyone with a grievance and a gun, we would be under constant attack from Sri Lankans, Basques, Tutsis, Irish, Chechens, Kashmiris, and, well, the list is endless. President Clinton is part of that cohort of American liberals who have been so marinated in cynicism about their own country that they find it difficult to discern evil in anyone else. These "internationalist" liberals criticize other nations only when they are allied with the United States (Israel, El Salvador, Chile), or when their sins are reminiscent of America's (South Africa).

And even at a moment when the U.S. was clearly the wounded party, clearly the victim (that most cherished of liberal categories), liberals like Bill Clinton were unable to switch gears and offer the United States total sympathy. In their universe, when a Third World people are arrayed against the United States, the former have the right of way.

In fact, the liberal tendency to dismiss or underestimate threats to the United States seems to be a permanent condition. Writing in (and seemingly on behalf of) the *New York Times*, reporter Robert F. Worth said this in February 2002: "As President Bush toured Asia last week, some world leaders worried publicly that the war on terrorism was starting to look suspiciously like the last great American campaign—against communism.... The language Mr. Bush and others have used to describe Al Qaeda terrorists sometimes sounds as though it could have been written by Cotton Mather."[74]

Worth was decrying what he saw as America's self-righteousness, a condition that has supposedly characterized the nation since colonial days. And though he agrees that the war on terror is necessary and that terrorists are evil, he is troubled by our tendency to think of ourselves as good. "To the extent that those enemies are seen as evil, America can regard itself as good, a desire rooted in the Puritan vision of establishing a new Eden in a fallen world."[75]

Christian doctrine to one side, the world is a fallen place—a roiling, corrupt, unstable, vicious, and unpredictable place—at least in many places. Absent American leadership and strength of mind during the twentieth century—it could have been infinitely worse. Liberal views, forged in Vietnam and tempered in Central America and beyond, got the world all wrong. Even worse, they got America all wrong.

# Epilogue

ON SEPTEMBER 11, 2001, THE *New York Times* carried an affectionate profile that must have seared the editors' consciences later in the day. On the front page of the "Arts and Culture" section, they published a sympathetic portrayal of two American terrorists who had been responsible for several bombings in the early 1970s. Bill Ayers and his wife, Bernadine Dohrn, were members of the Weather Underground— an extremely radical offshoot of the already radical New Left. The very first words of the piece were these by Ayers: "I don't regret setting bombs. I feel we didn't do enough."[1]

What they did was set off bombs in New York City's Police Headquarters (1970), the U.S. Capitol building (1971), and the Pentagon (1972), among other targets. Though he didn't kill anyone, Ayers told the *Times* that even today he finds "a certain eloquence to bombs, a poetry and a pattern from a safe distance."[2]

260 | USEFUL IDIOTS

The rest is the familiar boilerplate of ten thousand profiles of aging baby boomers—their children attending Ivy League colleges, their work for progressive causes, their guilt over material success, their struggles with monogamy. In anodyne phrases and soft lighting, the couple's cruel beginnings are domesticated and defanged. Would he do it again, Ayers is asked? He answers with poetry: "History says, Don't Hope/ On this side of the grave. / But then, once in a lifetime/ The longed-for tidal wave/ Of justice can rise up/ And hope and history rhyme."

This is repellent—not just the words of the terrorist (Osama bin Laden is reputedly quick with a verse as well), but the sickening openness to it by the nation's "newspaper of record." As it happens, this profile ran on a fateful day and is therefore recorded in italics forever. (Thousands of copies of that article were blasted into ashes at the World Trade Center.) These sorts of stories are characteristic not just of the *New York Times* but of liberal thinking as a whole. And what they prove is not just that liberals failed one of the two great moral tests of the twentieth century, but that they still do not know they failed and have not grappled with the implications of that failure.

The challenges posed by the two totalitarian systems of the twentieth century were comprehensive. They forced Americans to respond politically, culturally, economically, and above all morally. Liberals had no difficulty meeting the challenge of fascism. They found its barbarism despicable and said so. They feared its aggression. And they found its ideas disgusting. Even at the cost of 484,375 Americans (those who lost their lives in World War II), they were prepared to make whatever sacrifices were required to thwart and defeat that assault on world peace and simple decency.

Antifascism came as naturally to liberals as breathing—which is as it should be. The Nazis, in the more than half century since their defeat, have become synonymous with evil in our intellectual and cultural life—again, just as it should be. And yet the Communists, whose crimes were nearly comparable (and the debate about which was worse is shabby and irrelevant), have never even entered the evil category for liberals. Quick: try to think of a single movie about the horrors of Stalinism. This is not a failure of imagination. This is a moral meltdown.

Time and again throughout the latter part of the Cold War, liberals chose a morally perverse pose. They would seek to find any suspect motive or impure act on the part of the United States rather than confront the staggering scale of destruction and misery being wrought by our adversaries. They turned the New Testament wisdom about tending to the log in one's own eye before criticizing the mote in another's completely on its head. In the contest between the Soviet Union and the United States, it was the Soviets who had logs aplenty to notice; yet liberals could not see them because they couldn't avert their gaze from the motes in U.S. eyes.

This was more than perverse; it was also an intellectually flabby approach, since it offered the comforting illusion that the great threat facing the world was U.S. blundering and bullying rather than the true subversion and violence of the Communist world. Imagine for a moment that the skewed liberal version of reality were correct—that the great threats facing the world until 1991 were American militarism, support for "right-wing dictatorships," and supposed indifference to poverty. Wouldn't that be a comfortable place to live? Why, we could solve all of our problems with world town meetings and community initiatives! With just a bit of attitude adjustment on the part of Americans, racism, aggression, nuclear weapons, human rights abuses—all would disappear.

This solipsism was thus dangerous. If not for the fortitude of other Americans, it is very doubtful that the Cold War would have turned out as it did.

Writing in the *Wall Street Journal* on June 6, 1945, Thomas F. Woodlock saw the pattern that would persist for the next half century:

We all remember "Munich" and Neville Chamberlain's visit to the late unlamented Hitler. Munich has indeed become a symbol for "appeasement" and "appeasement" means a surrender of principle motivated by fear of the consequences entailed by sticking to it. What a lambasting the "liberals" gave poor Chamberlain in those days! How indignant they were at his surrender to Hitler merely on the grounds that Great Britain was in no position to engage the German military force!

... Today, we have a strikingly similar situation with a single difference, a difference, however, in the parties not in the situation.... And what do our "liberal" friends have to say as to this state of things? To those who in any form of words express criticism of Soviet Russia's course of action their answers largely boil down to a hurling of epithets such as "red-baiter," "Soviet hater," and "war-monger" and a final and supposedly crushing question, "Do you want war with Russia?"[3]

One of the most celebrated heroes of American history, Charles Lindbergh, saw his reputation shredded due to his failure to perceive the monumental evil of Nazism. Yet American liberals who committed the identical sin vis-à-vis the Communists—and demonstrate in ways small and large on an almost daily basis that they still do not get it—have paid no price in credibility for their appalling judgment. There is no scholar, public figure, diplomat, politician, or journalist who has suffered any diminution of reputation because he toadied to, appeased, justified, or excused the Communists. Not all liberals were "soft on communism"—a great, if overused phrase. Senator Henry Jackson of Washington State, New York's Daniel Patrick Moynihan (before he became a U.S. senator), Lane Kirkland of the AFL-CIO, and Martin Peretz of the *New Republic*, among others, distinguished themselves as vigorous and impassioned anticommunists. Still, *Great Liberal Cold Warriors* would make a short book.

Now that the United States is again faced with a threat from a determined and vicious enemy—Islamofascism—some liberals are clearly struggling to make sense of it. Decades of reflexive America-blaming have left them confused and dazed in the face of Islamic hatred. Journalist Christopher Hitchens, as hard a leftist as one is likely to encounter in polite society, ripped his colleagues for it, noting that the things the Islamofascists hate about the West are the very things—female emancipation, sexual license, religious tolerance—that the Left likes. Al Qaeda is not the militant wing of the antiglobalization movement.

That some precincts on the Left—even now—can find reasons to blame the United States for the hatred directed against it, is evidence that the rotten kernel of their appeasement and weakness throughout the second half of the Cold War was America-hatred. To explain the phenomenon would require a separate book—but in light of the sanguinary history of the world and the shining place America deserves on any list of humanitarian and civilized nations—it is a grotesque injustice as well as a sign of moral delinquency.

# Notes

## INTRODUCTION: NONE DARE CALL IT VICTORY

1. With apologies to T. S. Eliot.
2. Martin Gilbert, ed., *A History of the Twentieth Century, Volume III: 1952–1999* (New York: William Morrow & Co., 1999) 83.
3. Ibid., 674.
4. Address in Kiev, 1 August 1991, later mocked as the "Chicken Kiev" speech.
5. Winston Churchill's speech to Westminster College, 5 March 1946.
6. George H. W. Bush's speech at Yale University, broadcast on C-SPAN, 5 October 2001.
7. J. Robert Oppenheimer
8. Paul Hollander, *Anti-Americanism: Critiques at Home and Abroad, 1956–1990* (New York: Oxford University Press, 1992), liv.
9. Ibid., lvii.
10. Ibid., 462.
11. Julia Keller, "On the Record with Frances Fitzgerald," *Chicago Tribune*, 4 June 2000, 1.

12. Strobe Talbott, "Man of the Decade," *Time*, 1 January 1990.

13. Vaclav Havel's speech before U.S. Congress, 21 February 1990.

14. Lawrence Kaplan, "We're All Cold Warriors Now," *Wall Street Journal*, 18 January 2000, A26.

# CHAPTER ONE: THE BRIEF INTERLUDE OF UNANIMITY ON COMMUNISM

1. Ronald Reagan's speech to the National Association of Evangelicals, 9 March 1983.

2. Associated Press, 9 October 1983.

3. Bill Peterson, "Reagan's Use of Moral Language to Explain Policies Draws Fire," *Washington Post*, 23 March 1983, A15.

4. Steven Knott, "Reagan's Critics," *National Interest*, Summer 1996.

5. Juan Williams, "Writers for Speeches of President Claim Force Is With Him," *Washington Post*, 29 March 1983, A15.

6. Mary McGrory, "Experts and Citizens Speak Out, but Congress Turns Its Back; Fears," *Washington Post*, 10 November 1983, A3.

7. George Ball, "A Risky Mideast Course," *New York Times*, 21 November 1983, 23.

8. Anthony Lewis, "Onward Christian Soldiers," *New York Times*, 10 March 1983, A27.

9. Knott, *National Interest*.

10. Richard Pipes, *Communism: A History* (New York: Modern Library, 2001), 99. Steffens actually penned these words on the train traveling through Sweden before he had set foot in the USSR.

11. John Haynes, et al, *The Soviet World of American Communism* (New Haven: Yale University Press, 1998), 1.

12. George Stimpson, *A Book About American Politics*, (New York: Harper, 1952), 322.

13. Paul Johnson, *Modern Times: The World from the Twenties to the Eighties* (New York: HarperCollins, 1983), 360.

14. Vladmir Bukovsky, interview with author, 14 December 2001.

15. Sidney Hook, *Out of Step: An Unquiet Life in the 20th Century* (New York: HarperCollins, 1987), 320.

16. Ibid., 322.

17. Ibid., 367.

18. Hilton Kramer, *The Twilight of the Intellectuals* (Chicago: Ivan R. Dee, Inc., 1999), 80.

19. Peter Collier and David Horowitz, *Destructive Generation: Second Thoughts About the Sixties* (California: Touchstone Books, 1996), 167.

20. Robert Conquest, *Reflections on a Ravaged Century* (New York: W. W. Norton & Co., 1999), 136.

21. M. Stanton Evans, interview with author, 30 November 2001.

22. Ronald Radosh, "Truths and Excuses," *New Republic*, 3 January 2000.

23. Hook, 22.

## CHAPTER TWO: THE CONSENSUS UNRAVELS

1. Peter Rodman, *More Precious Than Peace: Fighting and Winning the Cold War in the Third World* (New York: Scribner, 1994), 106.

2. Podhoretz, Norman, *Why We Were in Vietnam* (New York: Simon & Schuster, 1984), 57.

3. Ibid., 55.

4. Ibid., 58

5. Paul Johnson, *A History of the American People* (New York: Harper Collins, 1998), 883.

6. Ibid., 81.

7. Ibid., 56.

8. Johnson depicted his 1964 opponent, Barry Goldwater, as a dangerous warmonger. This gave rise to a mordant joke among conservatives: "In 1964, they told me that if I voted for Goldwater there would be half a million men in Vietnam in six months. I did...and there were."

9. Rodman, 121–122.

10. Ibid., 123.

11. Paul Hollander, *The Survival of the Adversary Culture* (New York: Transaction Publications, 1991), 121.

12. Mitchell Stevens, Freedom Forum Online.

13. Johnson, *Modern Times*, 461.

14. John Corry, "Lessons of Vietnam are Explored by NBC," *New York Times*, 27 April 1985, 46.

15. Peter Braestrup, *Big Story* (New Haven: Yale University Press, 1983), 193.

16. Phil Cannella, interview by author, 17 August 2001.

17. "Notable Quotables," Media Research Center, 1997.

18. B. G. Burkett, *Stolen Valor: How the Vietnam Generation Was Robbed of Its Heroes and Its History* (Utah: Verity Press, 1998), 30–31.

19. Ibid., 31.

20. Braestrup, 39.

21. Burkett, 123.

22. Ibid., 126.

23. Ibid., 123.
24. Ibid., 118.
25. Ibid., 114.
26. Ibid., 258.
27. Ibid., 126.
28. Frum, David, *How We Got Here: The 70s—The Decade That Brought You Modern Life—For Better or For Worse* (New York: Basic Books, 2000), 308.
29. Richard Bernstein, "Susan Sontag: As Image and Herself," *New York Times*, 26 January 1989, 26.
30. Hollander, 271.
31. Paul Hollander, *Political Pilgrims: Western Intellectuals In Search of the Good Society* (New York: Transaction Pub, 1998), 198.
32. Hollander, *Political Pilgrims*, 267.
33. Ibid., 271.
34. Ibid.
35. Ibid.
36. Mary Therese McCarthy, *The Seventeenth Degree* (New York: Harcourt Brace, 1974), 215, 222.
37. Ibid., 230–231, 316.
38. Podhoretz, 99.
39. Jonathan Schell, *Observing the Nixon Years* (New York: Pantheon, 1989), 91.
40. Richard J. Whalen, *Taking Sides* (Boston: Houghton Mifflin, Inc., 1974), 177.
41. Henry Kissinger, *The White House Years*, (New York: Little Brown & Co., 1979), 512.
42. Ibid., 514.
43. Frank Gregorsky, *What's Wrong With Democratic Foreign Policy?* (A paper by the House Republican Study Committee, 1984), 4.
44. Burkett, 136.
45. Congressional Record, 25 March 1975, 8572.
46. Burkett, 135.
47. Ibid., 134.
48. Ibid., 135.
49. David Gates, "A Veteran Applies the Lessons of War," *Newsweek*, 7 May 1984, 18.
50. Hollander, 117.
51. Ibid.
52. Guenter Lewy, *America in Vietnam* (New York: Oxford University Press, 1978), 400–401.

53. Podhoretz, 103.
54. Kissinger, 444.
55. Frum, 305.
56. *New Republic*, "The Myths of Revolution," April 29, 1985.
57. Stephane Courtois, et. al, *The Black Book of Communism* (Cambridge: Harvard University Press, 1999), 572.
58. Guenter Lewy, "Seeing the Forest After the Fall," *Washington Times*, 27 December 1990, G4.
59. Charles Horner, "Who Won Vietnam?" *Commentary*, May 1994.
60. Tucker Carlson, "National Council of Castro Worshippers," *Weekly Standard*, 17 April 2000, 24.
61. Linda Ellerbee, CNN PrimeNews, 2 June 1989.
62. *Washington Post*, June 12, 1979.
63. Lynn Darling, "Joan Baez at 38," *Washington Post*, 29 June 1979, C1.
64. *New Republic*, "Myths of Revolution," April 29, 1985.

## CHAPTER THREE: THE BLOODBATH

1. Sydney Schanberg, "Cambodian Reds Are Uprooting Millions," *New York Times*, 19 April 1975.
2. John Barron and Anthony Paul, *Murder of a Gentle Land* (New York: McGraw-Hill, 1977), 26–28.
3. Courtois, 611.
4. Loung Ung, *First They Killed My Father* (New York: HarperCollins, 2000), 53.
5. Courtois, 611.
6. Ung, 103.
7. Courtois, 598.
8. Ung, 61.
9. Courtois, 599.
10. Kissinger, 518.
11. Courtois, 605.
12. Ibid., 605.
13. Ibid., 616.
14. Rodman, 186–187.
15. "Vietnam: Genocide," *Washington Post*, 12 October 1979, A14.
16. Courtois, 4.
17. Ibid.
18. Anthony Lewis, "Avoiding a Bloodbath," *New York Times*, 17 March 1975.
19. Rodman, 186.

20. Gregorsky, 16.
21. John Elvin, "Inside the Beltway," *Washington Times*, 17 April 1990, A6.
22. Ibid., A6.
23. Schanberg.
24. Sydney Schanberg, "*The Enigmatic Cambodian Insurgents*," New York Times, 13 March 1975, 1.
25. Schanberg, "Cambodian Reds."
26. Ibid., 15.
27. Schanberg, "Cambodian Reds."
28. Gregorsky, 16.
29. Anthony Lewis, "Abroad at Home," *New York Times*, 21 April 1975.
30. Charles Krauthammer, "Escape from Liberation," *Washington Post*, 19 April 1985, A27.
31. William Shawcross, *Side-show* (New York: Pocket Books, 1979), 389.
32. *The Killing Fields*, 1984.
33. Aric Press, "Kissinger's Fault?" *Newsweek*, 22 October 1979, 54.
34. Richard Cohen, "Kissinger Defends a Lie, and Forgets Kent State," *Washington Post*, 14 October 1979, B1.
35. Peter Rodman, "Rodman Responds," *American Spectator*, July 1981, 14.
36. Rodman, *More Precious Than Peace*, 184.
37. William Shawcross, "Shrugging Off Genocide," *Times of London*, 16 December 1994.

## CHAPTER FOUR: THE MOTHER OF ALL COMMUNISTS: AMERICAN LIBERALS AND SOVIET RUSSIA

1. Michael Barone, *Our Country* (New York: Free Press, 1990), 591.
2. Edward Walsh, "Carter Stresses Social Justice in Foreign Policy," *Washington Post*, 23 May 1977, A1.
3. Carl Gershman, "The Rise and Fall of the New Foreign Policy Establishment," *Commentary*, July 1980.
4. William F. Buckley, *Execution Eve* (New York: Putnam, 1975), 57, 58.
5. Hollander, *Political Pilgrims*, 296, 297.
6. Ibid., 307.
7. Courtois, 4.
8. Gregorsky, 12, 13.
9. Frum, 48.

10. David M. Alpern, "Inquest on Intelligence," *Newsweek*, 10 May 1976, 40.
11. Carl Gershman, "The Rise and Fall of the New Foreign Policy Establishment," *Commentary*, July 1980.
12. Ibid.
13. Jeanne Kirkpatrick, "Dictatorships and Double Standards," *Commentary*, November 1979, 40.
14. Ibid., 40.
15. Gregorsky.
16. Charles Martin, "Democrats and the Bomb," *Washington Post*, 17 April 1988, C5.
17. Jeanne Kirkpatrick, "Dictatorships and Double Standards," *Commentary*, November 1979.
18. After the breakup of the Soviet Union, the existence of stockpiles of chemical, biological, and nuclear weapons proved highly threatening to the West as they may have fallen into the hands of Islamic terrorists. This was the Soviet Union's posthumous "gift" to the West.
19. Thomas Firestone, "Four Sovietologists," *National Interest*, Winter 1988/89, 104.
20. "Notable Quotables," Media Research Center, 2 September 1991.
21. Ibid.
22. Ibid.
23. Christopher Gray, "Sixty-nine Years of 'Reform'," *Policy Review*, Fall 1986.
25. Ibid.
25. Conquest, 123.
26. Hollander, *Political Pilgrims*, 119.
27. Conquest, 123.
28. Hollander, *Political Pilgrims*, 113.
29. Hollander, *Political Pilgrims*, 115.
30. Saul Bellow, "Writers, Intellectuals, Politics," *National Interest*, Spring 1993.
31. Hook, 324.
32. Arnold Beichman, "Written in Stone on the Left Side," *Washington Times*, 8 December 1992, F4.
33. Henry Weinstein and Judy Pasternak, "I. F. Stone Dies," *Los Angeles Times*, 19 June 1989, 1.
34. Robert D. Novak, "I. F. Stone: Red and Dead," *Weekly Standard*, 22 June 1998, 16.
35. Larry King, "The Laws in the Land of Larry," *USA Today*, 14 August 1989, 2D.

36. "Media Watch," Media Research Center, July 1989.
37. Hollander, 12.
38. Richard Pipes, "The Evil of Banality," *New Republic*, 18 December 2000, 34.
39. Gray.
40. Courtois, 115.
41. Ibid., 131.
42. Hollander, *Political Pilgrims*, 161, 162.
43. Ibid., 164.
44. Conquest, 77.
45. Radosh, 39.
46. David Horowitz, *Radical Son* (New York: The Free Press, 1997), 73–74.
47. Edward Jay Epstein, *Dossier: The Secret History of Armand Hammer* (New York: Random House, 1996), 83–84.
48. Frank Gregorsky, "Is There Anything Left of Bipartisan Foreign Policy?" A report by the House Republican Study Committee, 45.
49. David Broder, "Nixon Wins Landslide Victory," *Washington Post*, 8 November 1972, A1.
50. Gregorsky, 11.
51. Carl Gershman, "The World According to Andrew Young," *Commentary*, August 1978, 18.
52. Ibid., 18.
53. Ibid., 19.
54. Courtois, 123, 124.
55. Ibid., 120.
56. Ibid., 120.
57. Ibid., 148.
58. Ibid., 150.
59. Ibid., 148.
60. Ibid., 154.
61. Gershman, "The World According to Andrew Young," 20.
62. An example: "What happens after the Soviet Union invades the Sahara Desert? *Answer*: Nothing for ten years and then a shortage of sand."
63. Richard Reiland, American Enterprise Online, 21 March 2001.
64. Ibid.
65. Conquest, 135.
66. Conquest, 106.
67. Margaret Shapiro, "Ex-Soviet Empire the New 'Sick Man of Europe'," *Washington Post*, 3 October 1992, A1.

68. R. Emmett Tyrrell, *The Liberal Crack-Up* (New York: Simon & Schuster, 1984), 157.
69. "The Cuban Presence, into the Angolan Vacuum," *Washington Post*, 1 February 1977, A17.
70. He changed his name to Natan Sharansky when he moved to Israel after serving ten years in the Gulag.
71. Kenneth Labich, "Andy Young, Dissident," *Newsweek*, 24 July 1978, 22.
72. Hollander, *Political Pilgrims*, 170.
73. Gray.
74. Ibid.
75. Jay Edward Epstein, "The Andropov File," *New Republic*, February 1983.
76. Ibid.
77. Ibid. 18
78. Epstein, "The Andropov File."
79. Gray.
80. Ibid.
81. Ibid.
82. Ibid.
83. Ibid.
84. "Notable Quotables," Media Research Center, 1990.
85. "Media Watch," Media Research Center, January 1990.
86. "And That's The Way It Isn't," a 1990 report by the Media Research Center, 126.
87. Ibid.
88. Gray.
89. Dinesh D'Souza, *Ronald Reagan* (New York: Free Press, 1997), 183.
90. Media Research Center, Notable Quotables.
91. D'Souza, 183.
92. Hollander, *The Survival,* 64, 65.
93. Media Research Center, Media Watch, January 1990, 5.
94. Michael Novak, "Pledging Allegiance," *Commentary*, December 1990, 62.
95. Reiland.
96. Paul Lewis, "Per Anger, 88, A Diplomat Who Helped Jews, Is Dead," *New York Times*, 29 August 2002, A21.
97. Lance Morrow, "Man of the Decade," *Time*, 1 January 1990, 42.
98. Strobe Talbott, "Man of the Decade," *Time*, 1 January 1990, 66.
99. Arch Puddington, "Voices in the Wilderness," *Policy Review*, Summer 1990.

100. Rodman, *More Precious Than Peace*, 290.
101. Reiland.
102. *Executive Intelligence Review*, vol. 20, no. 13.
103. Richard J. Barnet, "America Goes it Alone," *New York Times*, 23 October 1985, A23.

## CHAPTER FIVE: FEAR AND TREMBLING

1. D'Souza, 134.
2. Patrick Glynn, *Closing Pandora's Box* (New York: Basic Books, 1992), 327.
3. Gregorsky, 11.
4. George Weigel, "Shultz, Reagan, and the Revisionists," *Commentary*, August 1993, 50.
5. Norman Podhoretz, "Appeasement by any Other Name," *Commentary*, July 1983, 29.
6. Jay Winik, *On the Brink* (New York: Simon & Schuster, 1996), 282.
7. Ibid, 284.
8. David Corn, "Yuri, You Have Made Our Job Much Harder," *New York Times*, 9 September 1983, 19.
9. "To Persist with Arms Control," *New York Times*, 18 September 1983, Sec. 4, 18.
10. Glynn, 315.
11. James M. Markham, "Vast Crowds Hold Rallies in Europe Against U.S. Arms," *New York Times*, 23 October 1983, A1.
12. Ibid.
13. Winik, 212.
14. John Vinocur, "KGB Officers Try to Infiltrate Antiwar Groups," *New York Times*, 26 July 1983, A1.
15. Ibid., A1.
16. Ibid., A1.
17. Irving Kristol, "What's Wrong with NATO?" *New York Times Magazine*, 25 September 1983, 64.
18. Gabriel Schoenfeld, "A Charming Communist," *Commentary*, December 1995, 71.
19. Ibid.
20. Heritage Foundation Backgrounder No. 225, "The Hard Facts the Nuclear Freeze Ignores," 3 November 1982.
21. Glynn, 268.
22. Michael Fumento, "The Center for Defense Misinformation," *American Spectator*, April 1988.
23. Ibid.

24. Lloyd Grove, "A Liberal Dose of Idealism: Think Tank Fetes 30 Years on the Left," *Washington Post*, 4 October 1993, B1.
25. Glynn, 289.
26. Ibid., 304.
27. Schoenfeld, 70.
28. Ibid., 71.
29. Gregorsky, *What's Left of Bipartisan Foreign Policy?* (A paper by the House Republican Study Committee, 1988), 62.
30. "The Democrats on Defense," *Washington Post*, 10 April 1988, B6.
31. Gregorsky, *What's Left*, 72.
32. Ibid., 80.
33. "The Candidates Debate," *New York Times*, 12 October 1984, B4.
34. Christopher Madison, "Mondale Asking Voters if They Feel Safer," *National Journal*, 20 October 1984, 1960.
35. Gregorsky, *What's Left*, 83.
36. Ibid., 83.
37. Christopher Madison "Mondale Asking Voters if They Feel Safer," *National Journal*, 20 October 1984, 1960.
38. Jay Nordlinger, "Albright Then, Albright Now," *National Review*, 28 June 1999.
39. Gregorsky, *What's Wrong*, 17.
40. Michael Novak, "Arms and the Church, " *Commentary*, March 1982, 39.
41. Ibid., 39.
42. Novak, "Arms and the Church," 39.
43. Ibid., 41.
44. L. Bruce van Voorst, "The Churches and Nuclear Deterrence," *Foreign Affairs*, Spring 1983, 827.
45. Ibid., 827.
46. Judith Miller, "139 in Congress urge Nuclear Arms Freeze by U.S. and Moscow," *New York Times*, 11 March 1982, A1.
47. Heritage Foundation Backgrounder, No. 225, 7.
48. "U.S. Must Lead Elimination of A-Bombs," *Charleston Gazette*, 8 June 1998, 4A.
49. "Lutherans Ask Nuclear Ban," *New York Times*, 12 September 1982, 27.
50. Charles Austin, "2 Major Protestant Churches Call for End to Arms Race," *New York Times*, 18 December 1981, A26.
51. Briggs, Kenneth A., "Criticism of Reagan by Religious Leaders Rises," *New York Times*, 8 December 1981, A1.
52. Ibid., A1.

53. Gayle White, "Religion Q and A," *Atlanta Journal Constitution*, 18 February 1995, 6E.

54. Kathleen Teltsch, "Foundations Back Arms Control Plea," *New York Times*, 2 May 1982, A32.

55. Robert G. Kaufman, *Henry M. Jackson: A Life in Politics* (Seattle: University of Washington Press, 2000), 412.

56. Richard C. Gross, "Aspin Cautions Democratic Hopefuls on Nuke Freeze," UPI, 30 June 1983.

57. *MacNeill/Lehrer Newshour*, 16 January 1984.

58. Ibid.

59. "Artists Protest Nuclear Threat," *New York Times*, 11 December 1983, Sec. 1, 40.

60. Robert Macy, "438 Arrested At Test Site, Including Martin Sheen, Robert Blake," Associated Press, 6 February 1987.

61. "Freeze March; Rallying in New York," *Time*, 14 June 1982, 24.

62. Celestine Bohlen, "Soviets Formally Denounce Reagan's Joke," *Washington Post*, 16 August 1984, A32.

63. James Lardner, "The Bomb Schell," *Washington Post*, 22 April 1982, C1.

64. Ibid., C1.

65. TASS, "Pozner, Donahue Win Better World Society Award," 4 September 1987.

66. John Corry, "Week of Donahue Taped in Soviet Union," *New York Times*, 12 February 1987, C30.

67. MediaWatch, Media Research Center, June 1990, 6.

68. Ellen Goodman, "Samantha, The Kid Who Spoke for Us All," *Washington Post*, 3 September 1985, A15.

69. "Samantha to Query Presidential Hopefuls," UPI, 20 January 1984.

70. George W. Cornell, "Church Kids Shower Soviet Children With Fond Birthday Wishes," Associated Press, 27 May 1988.

71. "Soviet Dissident Reports New KGB Threat," *New York Times*, 31 August 1986, Sec. 1, 19.

72. Glynn, 4.

73. Ibid., 8.

74. Ibid., 9.

75. Glynn, 15.

76. Ibid., 292.

77. Glynn, 215, 216.

78. Ibid, 271.

79. Thomas Mahoney, "Disaster Lobby Gets Free Ride," *Air Conditioning, Heating, and Refrigeration News*, 16 October 1989, 26.

80. A. D. Horne, "Nuclear Climate More Lethal Than Predicted, Soviets Say," *Washington Post*, 9 December 1983, A44.

81. Ellen Goodman, "Do You Feel Safer?" *Washington Post*, 5 November 1983, A1.

82. Knott.

83. Judith Miller, "139 in Congress urge Nuclear Arms Freeze by U.S. and Moscow," *New York Times*, 11 March 1982, A1.

84. Fox Butterfield, "Anatomy of the Nuclear Protest," *New York Times*, 11 July 1982, Sec. 6, 14.

85. David Margolick, "Vote for Arms Talks," *New York Times*, 11 August 1982, A11.

86. Knott.

87. "26 Groups Join in Campaign for Arms Freeze," 17 October 1982, Sec. 1, 44.

88. Leslie Maitland, "Sources are cited for charge of Soviet Tie to Arms Freeze," *New York Times*, 13 November 1982, Sec. 1, 7.

89. Ibid.

90. Knott.

91. Judith Miller, "Congress Weighs Curb on Arms Race," *New York Times*, 21 March 1982, Sec. 1, 32.

92. Glynn, 321.

93. *Nightline* transcript #MP8037, "Day After Perils of Nuclear War."

94. James Litke, "Today Marks 40th Anniversary of Bulletin of Atomic Scientists," Associated Press, 12 December 1985.

95. Papers of the Presidents, University of Texas, Ronald Reagan's Address to the Nation, 23 March 1983.

96. Molly Ivins, "Why Is Bush So Starry-Eyed?" *Fort Worth Star Telegram*, 28 May 2000, 4.

97. Sean Vinck, "The Media War on Star Wars," *Weekly Standard*, 24 July 2000, 18.

98. Roberto Suro and Thomas E. Ricks, "More Doubts Are Raised on Missile Shield," *Washington Post*, 18 June 2000, A1.

99. Ibid.

100. Mary McGrory, "The Stars Spoke on Capitol Hill," *Washington Post*, 5 May 1988, A2.

101. Ibid.

102. George Wilson, "Senate Refuses to Slash 'Star Wars' Funding," *Washington Post*, 5 June 1985, A30.

103. Ibid.

104. Mark Steyn, "Mad About the Bomb," *The Spectator*, 28 July 2001, 20.

105. Ibid.

106. Tom Burgess, "Differing Viewpoints on SDI," *San Diego Union Tribune*, 22 October 1986, A10.

107. President Ronald Reagan's radio address, 13 July 1985.

108. 1984 White House summary of Pentagon SDI Document.

109. Charles Mohr, "Scientists Dubious Over Missile Plan," *New York Times*, 25 March 1983, A8.

110. Ibid., A8.

111. Ibid., A8.

112. Sean Vinck, "The Media on Star Wars," *Weekly Standard*, 24 July 2000, 18.

113. Insiders Report # 108, Heritage Foundation, December 1987, 4.

114. Vinck.

115. Gabriel Schoenfeld, "Way Out There in the Blue," *Commentary*, 1 May 2000, 75.

116. William J. Broad "The Nuclear Shield," *New York Times*, 30 June 2000, A1.

117. Gorbachev reportedly told Reagan at Reykjavik "Excuse me, Mr. President, but I do not take your idea of sharing SDI seriously." John J. Miller, "Fire in a Fake," *National Review*, 3 July 2000.

118. Charles Martin, "Democrats and the Bomb," *Washington Post*, 17 April 1988, C5.

119. Thomas Friedman, "Who's Crazy Here?" *New York Times*, 15 May 2001, A25.

120. D'Souza, 179.

121. Ronald Reagan, *An American Life* (New York: Simon & Schuster, 1990), 697.

122. John Donnelly, "Missile Test Success Raises Hope, Anxiety," *Boston Globe*, 16 July 2001, A1.

## CHAPTER SIX: EACH NEW COMMUNIST IS DIFFERENT

1. Radosh, 108, 109.

2. Gregorsky, *What's Wrong*, 12.

3. Armando Valladares, *Against All Hope* (New York: Knopf, 1986), 4.

4. Radosh, 123.

5. Hollander, *Political Pilgrims*, 236.

6. Hollander, *Political Pilgrims*, 240, 241.

7. Hollander, *The Survival of the Adversary Culture*, 249.

8. Ibid., 255.

9. Hollander, *Political Pilgrims*, 257.

10. Courtois, 651.

11. Hollander, *The Survival of the Adversary Culture*, 218.
12. Ibid., 652.
13. Radosh, 124.
14. Ibid., 127.
15. Hollander, *The Survival of Adversary Culture*, 251.
16. Valladares, 14.
17. Courtois, 651.
18. Ibid., 657.
19. Richard Grenier, "Fidel's Theme Park Going Out of Business," *Washington Times*, 23 February 1992, B4.
20. Ibid., B4.
21. Johnson, *Modern Times*, 628.
22. "Notable Quotables," Media Research Center, 23 December 1991, 5–6.
23. Ibid., 6.
24. Media Research Center, "And That's The Way It Isn't," 74.
25. Media Watch, Media Research Center, 1 May 1989.
26. ABC's *World News Tonight*, 3 April 1989.
27. "Cuba: Tourists Flock to Country Seeking Healthcare," *American Health Line*, 7 May 2001.
28. Cyber Alert, Media Research Center, 3 January 2000.
29. Ibid.
30. Ibid.
31. Ibid.
32. Nick Eberstadt, *The Poverty of Communism* (New Brunswick: Transaction Publishers, 1990) 199–206.
33. Hollander, *The Survival of the Adversary Culture*, 194.
34. John Podhoretz, "The Return of the Useful Idiots," *Weekly Standard*, 8 May 2000, 14.
35. L. Brent Bozell III, Statement at National Press Club, 9 May 2002.
36. Ibid.
37. Jay Nordlinger, "Who Cares About Cuba?" *National Review*, 11 June 2001.
38. Media Watch, Media Research Center, May 1989.
39. Jeanne Kirkpatrick, "Dictatorships and Double Standards," *Commentary*, November 1979, 44.
40. Courtois, 663.
41. Allen C. Brownfeld and J. Michael Waller, *The Revolution Lobby* (Washington, D.C.: Council for Inter-American Security: Inter-American Security Educational Institute, 1985), 94.
42. Jay Nordlinger, "In Castro's Corner," *National Review*, 6 March 2000.

43. Carlson, "The National Council of Castro Worshippers."

44. Ibid.

45. Rodman, *More Precious Than Peace*, 253.

46. John Goshko, "Militant Grenada: Marxist Regime Planned Force Up to 10,000 Strong," *Washington Post*, 17 December 1983, A1.

47. Rodman, *More Precious Than Peace*, 253.

48. Gregorsky, *What's Wrong*, 35.

49. Ibid., 35.

50. Ibid., 36.

51. Ibid., 36.

52. Richard Cohen, "War and Peace," *Washington Post*, 1 November 1983, C1.

53. Gregorsky, *What's Wrong*, 35.

54. Ibid., 37.

55. Ibid., 38.

56. "Patrick Leahy's Little Leak," *U.S. News and World Report*, 10 August 1987, 7.

57. Gregorsky, 39.

58. Robert Kaiser, "Is This a Foreign Policy or a Recipe for Disaster?" *Washington Post*, 30 October 1983, C1.

59. Rodman, *More Precious Than Peace*, 253.

60. Nordlinger, "In Castro's Corner."

61. Ibid.

62. Ibid.

63. Mark Starr, et al., "Sanctuary for Salvadorans," *Newsweek*, 11 July 1983, 27.

64. Media Research Center, "And That's The Way It Isn't," 148.

65. Rodman, *More Precious Than Peace*, 235.

66. TRB, "Reagan's Holy War," *New Republic*, 11 April 1983, 6.

67. Robert Kagan, *A Twilight Struggle* (New York: Free Press, 1996), 164.

68. Gregorsky, *What's Wrong*, 44.

69. Gregorsky, *What's Left*, 53.

70. Ibid., 70.

71. Ibid., 71.

72. Ibid., 75.

73. Ibid.

74. Ibid., 75.

75. Ibid., 55.

76. Gregorsky, *What's Wrong*, 15.

77. Rodman, *More Precious Than Peace*, 242.

78. Gregorsky, *What's Wrong*, 44.
79. Brownfeld and Waller, 43.
80. Tracy Early, "Many Churches Criticize US policy on El Salvador," *Christian Science Monitor*, 23 May 1983, 7.
81. Ibid., 44.
82. Brownfeld and Waller, 45.
83. Rodman, *More Precious Than Peace*, 244.
84. Brownfeld and Waller, 46.
85. Gregorsky, *What's Wrong*, 43.
86. Ibid., 34.
87. Gregorsky, *What's Wrong*, 32.
88. Gregorsky, *What's Left*, 76.
89. Ibid., 77.
90. Joshua Muravchik, *News Coverage of the Sandinista Revolution* (Washington, D.C.: American Enterprise Institute, 1988), 9.
91. Ibid.
92. Rodman, *More Precious Than Peace*, 231.
93. "The Challenge to Democracy in Central America," Department of State and Defense, 1986.
94. Rodman, *More Precious Than Peace*, 227.
95. Jeanne Kirkpatrick, "Dictatorships and Double Standards," *Commentary*, November 1979, 43.
96. Muravchik, 7.
97. Papers of the Presidents, televised speech by President Reagan, 9 May 1984.
98. Gregorsky, *What's Wrong*, 14.
99. Rodman, *More Precious Than Peace*, 228.
100. Ibid., 237.
101. Ronald Reagan's Address to the Nation, 16 March 1986.
102. Letters, *New Republic*, 14 April 1986, 4.
103. Letters, *New Republic*, 14 April 1986, 5.
104. "The Challenge to Democracy in Central America," Department of State, 32.
105. "The Situation in Nicaragua," U.S. Department of State Document, January 1986, 3.
106. "The Challenge to Democracy in Central America," Department of State, 31.
107. Courtois, 669.
108. "The Challenge to Democracy in Central America," 20.
109. Ronald Reagan's Address to the Nation, 16 March 1986.
110. Ibid., 25.

111. Courtois, 672.
112. "Nicaraguan Refugee Update," White House Document, March 1986.
113. Hollander, *The Survival of the Adversary Culture*, 256.
114. Ibid., 255.
115. Radosh, 184.
116. Hollander, *The Survival of the Adversary Culture*, 258.
117. Radosh, 185.
118. Ibid., 177.
119. Muravchik, 50.
120. Ibid., 58.
121. Ibid., 58.
122. Notable Quotables, Media Research Center, 7 January 1991.
123. Hollander, *The Survival of the Adversary Culture*, 223.
124. Muravchik, 30.
125. Ibid., 196.
126. Hollander, *Anti-Americanism: Critiques at Home and Abroad* (New York: Oxford University Press, 1992), 129.
127. Gregorsky, *What's Wrong*, 14.
128. Gregorsky, *What's Left*, 81.
129. Ibid., 76.
130. Ibid., 76.
131. *White House Digest*, 20 June 1984, 4–5.
132. Rodman, *More Precious Than Peace*, 242.
133. Ibid., 242–243.
134. Brownfeld and Waller, 102.
135. Ibid., 102.
136. Jay Winik, *On the Brink* (New York: Simon & Schuster, 1996), 563.
137. Charles Babcock, "Nothing 'Improper' Found in Intercepts, Boren Says," *Washington Post*, 5 October 1991, A8.
138. Winik, 563.
139. Media Research Center, "And That's The Way It Isn't," 140.
140. Ibid., 144.
141. Gregorsky, *What's Left*, 76.
142. Hollander, *The Survival of the Adversary Culture*, 199.
143. Hollander, *Anti-Americanism*, 128.
144. Ibid., 129.
145. Hollander, *Anti-Americanism*, 106.
146. *Congressional Record*, 28 April 1983, H 24439.
147. "Democrats' Response to Reagan's Speech," Facts on File, *World News Digest*, 29 April 1983, 303 F1.
148. Ibid.

149. "The Case for the Contras," *New Republic*, 24 March 1986, 7.
150. Brownfeld and Waller, 125.
151. Rodman, *More Precious Than Peace*, 241.
152. "Ortega Visit Spurs New Thought on Contra Aid," UPI, 6 May 1985.
153. Rodman, *More Precious Than Peace*, 401.
154. Rodman, *More Precious Than Peace*, 434.
155. "The Nicaragua Obsession," *New York Times*, 25 February 1990, Sec. 4, 18.
156. Notable Quotables, Media Research Center, 24 December 1990.
157. Ibid.
158. Rodman, *More Precious Than Peace*, 441.
159. "Appointments Insult Human Rights Cause," *National Catholic Reporter*, 10 August 2001, 32.
160. Steve Inskeep, *All Things Considered*, National Public Radio, 10 July 2001.
161. Ibid.
162. Ibid.

## Chapter Seven: Post-Communist Blues

1. "MediaWatch," Media Research Center, October 1990, 5.
2. "Notable Quotables," Media Research Center, 24 December 1990.
3. Ibid.
4. "Notable Quotables," Media Research Center, 1992.
5. Ibid.
6. "Notable Quotables," Media Research Center, 24 December 1990.
7. Ibid.
8. "Notable Quotables," Media Research Center, 21 January 1991.
9. "MediaWatch," Media Research Center, June 1990.
10. "Notable Quotables," Media Research Center, 21 January 1991.
11. Ibid.
12. Ibid.
13. "Notable Quotables," Media Research Center, January 1994.
14. "Notable Quotables," Media Research Center, 23 December 1991.
15. "Notable Quotables," Media Research Center, 1990.
16. John O'Sullivan, "A Boy and a Nation," *National Review*, 21 February 2000.
17. John Podhoretz, "The Return."
18. Special Report, "Back to the 'Peaceable Paradise,'" Media Research Center, 23 May 2000.
19. "Communism Still Looms As Evil to Miami Cubans," *New York Times*, 10 April 2000, A1.

20. Special Report, "Back to the 'Peaceable Paradise,'" Media Research Center, 23 May 2000.
21. Ibid.
22. Thomas Sowell, "A Question Yet to Be Asked," *Washington Times*, 12 February 2000, A12.
23. Nordlinger, "In Castro's Corner."
24. Special Report, "Back to the Peaceable Paradise," Media Research Center, 23 May 2000.
25. Ibid.
26. Ibid.
27. Ibid.
28. Mona Charen, "Cold War Reminder," *Washington Times*, 3 February 2000.
29. Special Report, "Back to the Peaceable Paradise?" Media Research Center, 23 May 2000.
30. Ibid.
31. Ibid.
32. Ibid.
33. Deroy Murdock, "Jackboot Janet Stomps NBC News," *Washington Times*, 14 May 2000, B5.
34. Victorino Matus, "The Media Mob vs. Cuban Americans," *Weekly Standard*, 8 May 2000.
35. Ibid.
36. Ibid.
37. Special Report, "Return to Peaceable Paradise" Media Research Center, 23 May 2000.
38. Peter Beinart, "TRB from Washington,"*New Republic*, 20 September 2001.
39. Katha Pollitt, "Put Out No Flags," *The Nation*, 8 October 2001.
40. Ross Douthat, "Kumbaya Watch," *National Review*, 26 September 2001.
41. Mona Charen, "America Hatred Dies Hard," *Indianapolis Star*, 29 September 2001.
42. Douthat.
43. Ibid.
44. "Chattering Asses: Village Voice Edition," *Weekly Standard*, 15 October 2001.
45. John Leo, "Seeing no Evil," *Washington Times*, 16 October 2001.
46. "The Surprisingly Good Guys List, II," *Weekly Standard*, 29 October 2001, 4.

47. Ibid.
48. Douthat.
49. Noam Chomsky, *9-11* (New York: Seven Stories Press, 2001).
50. Douthat.
51. Ibid.
52. Leon Fuerth, "Digging Out," *Washington Post*, 16 September 2001, B7.
53. Notable Quotables, Media Research Center, 24 December 2001.
54. Terry Teachout, "Prime Time Patriotism," *Commentary*, November 2001, 52.
55. Douthat.
56. "Notebook," *New Republic*, 8 October 2001, 10.
57. "At War IV: Hall of Shame," *National Review*, 15 October 2001.
58. *Andrewsullivan.com*, 18 September 2001.
59. Douthat.
60. Ibid.
61. "Chattering Asses, Post-Grad Division," *Weekly Standard*, 8 October 2001.
62. Paul Hollander, "Anti-Americanism Revisited," *Weekly Standard*, 22 October 2001, 23.
63. Paul Hollander, "It's a Crime that Some Don't See This as Hate," *Washington Post*, 28 October 2001, B3.
64. Ibid.
65. Alison Hornstein, "My Turn," *Newsweek*, 17 December 2001, 14.
66. Noemie Emery, "The Crybaby Left," *Weekly Standard*, 17 December 2001, 25.
67. Noemie Emery, "Look Who's Waving the Flag Now," *Weekly Standard*, 15 October 2001.
68. Ibid.
69. Paul Craig Roberts, "Setbacks on the Home Front," *Washington Times*, 1 November 2001.
70. Ibid.
71. Emery, "The Crybaby Left."
72. Roberts.
73. Wesley Pruden, "A Clinton Chorus of 'America the Ugly,'" *Washington Times*, 9 November 2001, A4.
74. Robert F. Worth, "Truth, Right, and the American Way," The Week in Review, *New York Times*, 24 February 2002.
75. Ibid.

## Epilogue

1. Dinitia Smith, "No Regrets for a Love of Explosives," *New York Times*, 11 September 2001, E1.
2. Ibid.
3. Thomas F. Woodlock, *Wall Street Journal*, 6 June 1945.

# Acknowledgments

THIS BOOK OWES ITS EXISTENCE TO A thunderstorm—a series of storms really—that made eastbound air travel impossible one afternoon in the winter of 2000. Stranded at the Indianapolis airport, I was incredibly lucky to run into Herb London, the President of the Hudson Institute and one of the most engaging and intelligent fellow strandees one could ask for. With our coffee we chewed over the Cold War and liberalism, and Herb encouraged me to translate my idle musings into a book. Without that thunderstorm, Herb London, and the Hudson Institute, the book you hold in your hands would not have been written.

I am also indebted to Ambassador Curt Winsor and the Donner Foundation as well as to Michael Joyce and the Bradley Foundation for their support of and enthusiasm for this project. I'm grateful to my agent, Carole Mann, for her expertise. Ken Weinstein of the Hudson Institute was unfailingly helpful. My editor, Bernadette Malone, was sheer pleasure to work with and the book significantly benefited from her keen eye. Al Regnery was patient through a series of crises that delayed delivery of the manuscript and offered extremely helpful suggestions and edits based on his deep knowledge of the subject.

Barbara and Michael Ledeen opened their files and their hearts to me, for which I'm eternally grateful. Bob Andrews offered a stream of e-mails containing little-known facts, quotes, and tidbits of history. Ramesh Ponnuru and Kate O'Beirne shared their encyclopedic knowledge, as did Elliott Abrams, Karlyn Bowman, Jay and Lyric Winik, John Lenczowski, B. G. Burkett, Josh Gilder, Vladimir Bukovsky, Arnold Beichman, Peter Rodman, Brent Bozell, and Brent Baker. Though I wish I could take credit, it was Peter Collier who originated the idea for the title.

I cannot begin to name all of the scholars, historians, and journalists whose work made mine easier, but I must single out Frank Gregorsky, my old colleague from the Reagan White House, who exhaustively documented the statements of liberal Democrats on foreign policy for the better part of two decades, and who generously supplied the fruit of his careful research.

Danielle Frum and Melinda Sidak were kind enough to read early drafts and offer helpful comments and ideas. Linda Garner kept me from pulling out my hair over computer problems. David Limbaugh offered terrific moral support. And Priscilla Buckley cheerfully offered the perspective of a lifelong Cold Warrior. But then, Priscilla does everything cheerfully!

My assistant Jeanne Massey kept everything organized: the book files, my notes, the "book books," my other professional obligations, and much more besides. Jeanne is the human equivalent of Lexis-Nexis in my life, so easy does she make things. Laura Jones's unfailing reliability with the children and all around competence made it possible for me to focus on events twenty and thirty years old confident that more pressing matters would not be neglected.

Finally, I wish to thank my husband, Robert Parker, who instinctively distrusts praise, but deserves more than I can express.

—MONA CHAREN
*Great Falls, Virginia*

# Index

ABC News, 52, 86, 161, 165; Nicaragua and, 219, 227; September 11 and, 251

Abernethy, Bob, 112–13, 236

ABM Treaty, 130–31, 164, 165

Abrams, Elliott, 194, 228

ACLU. *See* American Civil Liberties Union

Acton, Lord, 231

Adams, Eddie, 32–33

Addams, Jane, 141

Afghanistan: Communism in, 81, 155, 197; Soviet invasion of, 77, 84

Africa, 81

Agency for International Development, 250

Albania, 233–34

Albright, Madeleine, 121, 136

Alexander, Bill, 218

Alexander, Jane, 160–61

Ali, Muhammad, 144

Allen, Woody, 20

"America: A Tribute to Heroes" telethon, 251

*America in Vietnam* (Lewy), 47

American Baptist Churches, 141

American Bar Association, 159

American Civil Liberties Union (ACLU), 89, 160, 212

American Friends Service Committee, 140

American Jewish Congress, 159

American Left. *See* American liberals

American liberals: ABM Treaty and, 165; anti-Americanism of, 29; antifascism and, 260–61;

American liberals *(continued)*
  arms control and, 129, 132–37,
    141–45; Cambodia and, 65–69;
    Cold War and, 1, 4–10, 14, 231,
    261; Communism and, 10,
    14–15, 77–86, 96–97, 117; Cuba
    and, 173–77; Elian Gonzalez
    controversy and, 231, 236–45; El
    Salvador and, 193–96; Grenada
    and, 189; Nicaragua and, 206,
    212–18, 222–27; nuclear war
    and, 155–58; Reagan and,
    119–23; revisionism and, 230;
    Sandinista Liberation Front and,
    212–18; September 11 and, 231,
    245–57; Soviet Union and,
    11–15, 77–117, 171, 231;
    Vietnam War and, 27–31, 38,
    43–44; Watergate and, 48. *See
    also* United States
American Lutheran Church, 140
American Public Health Association,
  159
Americans for Democratic Action,
  141
American-Soviet Friendship
  Committee, 88
Americas Watch, 194
Amherst College, 12
Anchorage, Alaska, 124
Anderson, John Ward, 243
Andropov, Yuri, 143, 148, 150–51;
  arms control and, 123; Soviet
  leadership and, 106–9
Angkar (Organization), 58, 60, 61,
  63–64
Angola, 81, 105, 155
Anspach, Susan, 212
Antigúa, 188
"Apes on a Treadmill" (Warnke),
  82–83
appeasement, 16, 142, 154, 231,
  261–62
Arbatov, Georgi, 130
Arias, Oscar, 219–20
Arias plan, 224

Aristotle, 90
Arms Control and Disarmament
  Agency, 82
arms control: ABM Treaty and,
  130–31; American liberals and,
  129, 132–37, 141–45; Catholic
  Church and, 137–39; Korean
  airliner incident and, 125;
  Protestant churches and, 139–41;
  SDI and, 162–70; Soviet Union
  and, 11–12, 119–51, 124–31,
  131–37, 154–55; war and,
  151–52; World War I and, 151.
  *See also* nuclear war
arms race. *See* arms control
Armstrong, James, 140, 177
Arnett, Peter, 33
Asia, 54
Asner, Ed, 144, 212
Aspin, Les, 136, 142–43
Associated Press, 11, 32–33
Astorga, Nora, 212–13
AuCoin, Les, 136
Austria, 3
Avila, Jim, 238
Ayers, Bill, 259–60

Baader-Meinhof gang, 85
Baez, Joan: Vietnam War and, 52–54
Baghdad, Iraq, 33
Baker, James, 5
Baldwin, Hanson, 36–37
Ball, George W., 13
Baltic states, 3, 5
Baltodano, Prudencio, 210
Bangkok, Thailand, 67
Barash, David P., 254
Barbados, 188
Barner, Richard, 46
Barnes, Michael, 217, 218
Barnet, Richard J., 116
Barron, John, 57–58
Baruch-Lilienthal plan, 19
Batista, Fulgencio, 173, 179, 183
Bay of Pigs, 187
Belafonte, Harry, 144

Belarus, 5, 17
Belasco, David, 94
Bellow, Saul, 23
Ben Tre, 33
Berger, Sandy, 10
Berlin Wall: fall of, 2–4, 9
Bernstein, Leonard, 4
Berthold, Richard A., 255
Bessmertnykh, Alexsandr, 116
Better World Society, 147
Bialer, Seweryn, 13–14, 105
Big Lie, 37
bin Laden, Osama, 247, 255, 260
Bishop, Maurice, 188, 192, 230
Bitterman, Jim, 234
*Black Book of Communism* (Courtois), 60, 91, 101, 102
Blake, Robert, 144
Blumberg, Abraham, 209
Blumenthal, Sydney, 113
Boell, Heinrich, 220
Boland Amendment, 225
Boland, Edward, 218
Bonior, David, 197, 202, 218, 220
Booth, Cathy, 180
Borge, Tomas, 206, 207–8, 210, 214
Bosnia, 250
*Boston Globe*, 170
Boxer, Barbara, 164, 190
Bozell, Brent, 184
Bradley, Bill, 10, 121
Bragg, Rick, 244
Brandenburg Steel Mill, 232
Brandt, Willy, 127
Brazil, 177
Brezhnev, Leonid, 5; Carter and, 78; Communism and, 155; detente and, 133; fear of nuclear war and, 146; Politburo of, 107; Soviet leadership and, 110
Britain: nuclear war and, 154; World War I and, 152; World War II and, 17
Broad Opposition Front, 207
Broder, David, 79–180
Broton, Elizabet, 237, 238

Brown, Amos, 252–53
Brown, Chester L., 33
Brown, George, 199–200
Brown, Harold, 132
Browne, Jackson, 212
Brown University, 254
Brzezinski, Zbigniew, 82
Buckley, William, Jr., 36, 78, 161
Budapest, Hungary, 3
Bukovsky, Vladimir, 84, 91
Bulgaria, 4, 233
Bulletin of Atomic Scientists, 161
Bundy, McGeorge, 137, 223
Burkett, B. G.: Vietnam War and, 32, 34–36, 37
Burns, John F., 106, 109
Bush, George H. W., 134, 250; Cold War and, 2, 5, 6
Bush, George W., 228, 240, 257; September 11 and, 252–54
"Butcher of the Ukraine." *See* Khrushchev, Nikita

Cairy, Eric, 188
Caldicott, Helen, 145–46, 158–59
Calero, Adolfo, 223
California, 143, 192, 200
Cambodia, 42; American liberals and, 65–69; collective agriculture and, 58–59, 102; Communism in, 55–64, 81, 155; Democratic Party and, 66; Khmer Rouge and, 65–69; terror-famine in, 58–63; United Nations and, 70; U.S. and, 69–76;
Campbell, Joan, 141, 238
Campbell-Bannerman, Sir Henry, 152
Campus Crusade for Christ, 210
Canada, 122
Cannella, Phil, 33
capitalism: Communism and, 234–35; Great Depression and, 15
Cardinal, Ernesto, 212
Carey, Mariah, 251
"Carlos." *See* Ramirez-Sanchez, Ilyich
Carlson, Tucker, 187

Carmichael, Stokely, 28

Carr, Bob, 66

Carrington, Lord, 11–12

Carter, Jimmy, 12, 132; arms control and, 153; Cold War and, 77; Communism and, 81–86; Communism in, 96–97; El Salvador and, 195; Nicaragua and, 205–6, 207; nuclear war and, 150

Castro, Fidel, 89, 97, 141; attempts to assassinate, 24, 80; Clinton and, 236, 239–40; Communism and, 245; Communism in Cuba and, 171–87; Elian Gonzalez controversy and, 236–41; Marxism of, 173; Nicaragua and, 187, 206; press and, 243–44; Soviet Union and, 85. *See also* Cuba

Catholic Church: arms control and, 137–39; El Salvador and, 200–201; Nicaragua and, 209–10; Poland and, 3; Somoza and, 209–10

CBS, 36, 52, 111, 145, 219

CDI. *See* Center for Defense Information

Center for Defense Information (CDI), 130, 131

Center for International Policy, 230

Central America: Reagan and, 198–204, 216–18, 229; U.S. and, 196, 198–204

Central Intelligence Agency (CIA), 80, 223–24

Central Michigan University, 255

Chamberlain, Neville, 16, 261–62

Chambers, Whittaker, 15, 20

Chamorro, Violetta, 218, 227

Chancellor, John, 236

Cheka, 93, 96

Chernenko, Constantin, 109–10

Chesimard, Joanne. *See* Shakur, Assata

Chile, 177, 194, 197, 218, 256

China, 81; collective agriculture and, 58, 102; Communism in, 7, 24, 52, 77–78, 171; criticism of, 14; Middle East and, 137; opening to, 77–78; ping pong diplomacy and, 78; Vietnam War and, 54

Chomsky, Noam, 47, 212, 248–49

Christian, Shirley, 215

Christian Coalition, 160

Christopher, Warren, 205

*Chronicle of Higher Education*, 254

Chung, Connie, 233

Church, Frank, 79–80

Churchill, Winston, 6, 18

Church of the Brethren, 200

CIA. *See* Central Intelligence Agency

CISPES. *See* Committees in Solidarity with the People of El Salvador

Citizens Against Nuclear War, 159

Citizens Party, 141

civil rights, 94, 98, 141

Clark, Ramsay, 40

Cleary, Paul, 51–52

Clifford, Clark, 223

Clift, Eleanor, 242

Clinton, Bill, 113, 121; Castro and, 236, 239–40; Cold War and, 9–10; Elian Gonzalez controversy and, 242; September 11 and, 256–57

Clinton, Hillary, 29, 142, 249

*Closing Pandora's Box* (Glynn), 132, 151

CNN, 33, 111, 184–85, 219

Coffin, William Sloane, Jr., 40, 140, 212, 220–21

Cohen, Richard, 73, 190, 238

Cohen, Stephen, 92, 106–7, 110–11

Colby, William, 159

Cold War: American liberals and, 1, 4–10, 14, 231, 261; cause of, 6; conclusion of, 1–2; end of, 2–4; foreign policy and, 10; Gorbachev and, 6–7; nuclear war and, 4, 7; origins of, 15–22; revisionism and, 1, 4–10, 18–19;

Stalin and, 6; victory in, 1–10;
World War II and, 2. *See also*
Communism
Coles, Robert, 209
collective agriculture: Cambodia and,
58–59; Soviet Union and, 98–105
Collier, Peter, 21
Colombia, 180
Colorado, 196
Columbia University, 13, 29, 105,
173
Comintern, 202
Commager, Henry Steele, 12
*Commentary* magazine, 81, 139
*Commies* (Radosh), 171
Committee for Cultural Freedom, 90
Committee for Peace in a Nuclear
Age, 141
Committees in Solidarity with the
People of El Salvador (CISPES),
194
Communism: American liberals and,
10, 14–15, 77–86, 96–97, 117;
anti-, 77–86; capitalism and,
234–35; collapse of, 2–4;
Democratic Party and, 24, 80;
Gorbachev and, 2–3, 5; Great
Depression and, 29; human rights
and, 84–85; nostalgia for,
231–36; nuclear war and,
157–59; opposition to, 1; reality
of, 9; reform of, 2–3, 5; Third
World and, 122; threat of,
114–15; Vietnam War and,
29–30. *See also* Cold War; Soviet
Union
Communist Party USA, 15, 141
Congressional Black Caucus, 186
Connecticut, 199
Conquest, Robert, 91
Contadora group, 192
Contadora Process, 224
containment, 19, 81
Conte, Silvio, 112
Contras, 221–27
Contreras, Joseph, 241

Conyers, John, 192
Corn, David, 125
Cornell University, 157
Cortazar, Julio, 220
Council for a Nuclear Weapons
Freeze, 141
Council on Foreign Relations, 10
Couric, Katie, 180, 239
Cranston, Alan, 121, 143, 161, 200,
201
Crimea, 95
Crockett, George, 134, 200
Cronkite, Walter, 33–34, 38–39, 195
Crosby, David, 44
*Crucible, The* (Miller), 20
Cruise, Tom, 251
Cruz, Arturo, 223
Cuba, 107; American liberals and,
173–77; arms control and, 126;
Castro and, 171–87; Communism
in, 7, 172–87; El Salvador and,
197; Grenada and, 188; human
rights and, 185; Nicaragua and,
216; political persecution in,
177–78, 186–87; political
pilgrims and, 172. *See also*
Castro, Fidel
Cuban Democratic Workers'
Confederation, 185
Cuban Missile Crisis, 26, 122, 155
"Cuban Solzhenitsyn." *See*
Valladares, Armando
Czechoslovakia, 3–4, 8, 199

Davies, Joseph, 93–94
Davis, Angela, 21, 174
*Day After, The*, 161
Dellinger, David, 175
Dellums, Ron, 43–44, 133, 192, 193
Democratic National Committee, 193
Democratic Party: Cambodia and, 66;
Communism and, 24, 80;
democracy and, 195; nuclear war
and, 160; Vietnam War and, 30,
42. *See also* American liberals
Democratic Socialists of America, 215

Deng Xioping, 109
Dershowitz, Alan, 244
detente, 78, 120, 133
deterrence, 162
Dewey, John, 90
Dewey Canyon III, 45–46
Dewey Commission, 20
Dewhurst, Colleen, 144
de Young, Karen, 206, 214
Dilley, Russell, 177
diplomacy: megaphone, 12; ping pong, 78
*Dissent* magazine, 212
Dith Pran, 64
Dixon, Julian, 192
DKP. *See* German Communist Party
Dobrynin, Anatoly, 129, 133, 169
Doctorow, E. L., 8
Dodd, Christopher, 66, 199, 218, 222–23, 229
Doder, Dusko, 107, 112
Dohrn, Bernadine, 259–60
Dominica, 188
Dominican Republic, 180
*Donahue*, 147, 148, 165
Donahue, Phil, 146–47
Dorgan, Byron, 203
*Dossier: The Secret History of Armand Hammer* (Epstein), 96
Dost Mohammad, 57
Douglas, Michael, 212
Downey, Tom, 44, 66
Dreiser, Theodore, 88
Duarte, Jose Napoleon, 204
DuBois, W. E. B., 106
Dudman, Richard, 71–72
Dukakis, Michael, 83, 168
Duke University, 255
Duranty, Walter, 15, 16, 87, 88
Dymally, Mervyn, 192
Dzerzhinsky, Feliks, 92–93, 96

*Early Show*, 241
Eberstadt, Nicholas, 183
Egypt, 84
Ehrlich, Paul, 157–58

Eisenhower, Dwight, 24, 30
elections: in Czechoslovakia, 3; in Hungary, 3; in Nicaragua, 217
Ellerbee, Linda, 52
El Salvador, 218, 256; American liberals and, 193–96; antiwar movement and, 47; Catholic Church and, 200–201; Communism in, 85, 187, 193–204; death squads in, 196; democracy and, 197; FMLN and, 197, 199; Sandinista Liberation Front and, 202–3; U.S. and, 227
encirclement, 81
Episcopal Church, 140, 247–48
Epstein, Edward Jay, 96, 108
Ethiopia: collective agriculture and, 102; Communism in, 81, 155
Europe: Communism in, 3; nuclear war and, 154
*Evening News*, 145, 219, 232, 241

"Facing Up to the Bomb," 161
Fainsod, Merle, 91
Fairlie, Henry, 209
Fair Play for Cuba Committee, 173
Falk, Richard, 212
Falwell, Jerry, 246
Farabundo Marti National Liberation Front. *See* FMLN
Farrell, Mike, 212
*Fate of the Earth, The* (Schell), 145, 146
Fazio, Vic, 224
FBI. *See* Federal Bureau of Investigation
Federal Bureau of Investigation (FBI), 80
Federal Council of Churches, 139
Feffer, Itzhak, 94–95
Feiffer, Jules, 144
Ferdinand, Franz, 151
Ferraro, Geraldine, 134–35
*Fire in the Lake* (Fitzgerald), 8, 53
"Firestorm" (Morris), 144

First Evangelical Church of Central America, 210

*First They Killed My Father* (Loung Ung), 58

Fitzgerald, Frances, 8, 41–42, 53

FMLN (Farabundo Marti National Liberation Front), 187; Communism in El Salvador and, 193–95, 197, 199

Fonda, Jane: arms control and, 144; Vietnam War and, 44, 46, 53–54

Foner, Eric, 254–55

Ford, Gerald, 65–66

*Foreign Affairs*, 132

foreign policy: civil rights and, 98; Cold War and, 10

*Foreign Policy* magazine, 80

Foster, William Z., 15

Four-Power Agreement, 19

France: Khmer Rouge and, 63; nuclear war and, 154; Vietnam War and, 53

Frank, Barney, 196–97

Frederick the Great, 224

French Action Directe, 85

Frente Sandinista de Liberación Nacional (FSLN), 207

Friedan, Betty, 212

Friedman, Thomas, 168–69, 245

Friends of the Earth, 159

*Front, The*, 20

Frost, David, 73

Frum, David, 48–49

FSLN (Frente Sandinista de Liberación Nacional), 207

Fuentes, Carlos, 220

Fuerth, Leon, 250

Galbraith, John Kenneth, 78–79, 105

Garcia, Robert, 202, 218

Garcia Marquez, Gabriel, 220

Geneva Convention, 29, 67

Georgetown University, 139, 256

Gere, Richard, 253

German Communist Party (DKP), 127–28

Germany: arms control and, 151; Communism in, 3–4; freedom and, 3; Nazi, 90; World War I and, 151–53; World War II and, 19

Geyelin, Phillip, 164

Geyer, Georgie Anne, 199

Ginzburg, Alexander, 84, 86, 106

Gitlin, Todd, 174

Gizbert, Richard, 235

*glasnost*, 5, 110

Glynn, Patrick, 132, 151–53

Gonzalez, Elian, 181–82; American liberals and, 231, 236–45; Castro and, 236–41; press and, 241–45

Gonzalez, Henry, 192

Gonzalez, Juan Miguel, 236, 237–38, 239, 240

Gonzalez, Lazaro, 236, 237

Gonzalez Bridon, Jose Orlando, 185

Goodfellow, Bill, 230

Goodman, Ellen, 148–49, 158

Gorbachev, Mikhael: arms control and, 131; Cold War and, 6–7; Communism and, 2–3, 5; Cuba and, 185; reforms of, 5; SDI and, 116, 167, 169–70; Soviet leadership and, 109, 110–15

Gordon, Michael R., 163

Gore, Al, 160, 164–65, 250

Gore, Tipper, 242

Gornick, Vivian, 247

Gott, Richard, 75

GPU, 93, 102

Grass, Günther, 220

Great Depression, 15, 29

Grechko, Marshal, 165

Greece, 19, 80

Greene, Graham, 220

Gregory, Dick, 44

Grenada: American liberals and, 189; Communism in, 81, 85, 155, 187–93; Cuba and, 188; political pilgrims and, 172; Reagan and, 189–92; Soviet Union and, 188; U.S. and, 188–93, 227

Grenier, Richard, 179–80
Gromyko, Andrei, 124, 127
Ground Zero, 141
Guevara, Che, 186; Communism in
    Cuba and, 175–77
Gulag Archipelago, 84
Gulf War, 33
Gumbel, Bryant, 180, 241
*Guns of August, The* (Tuchman), 152
Guthrie, Arlo, 127

Haing Ngor, 64
Haiphong, Vietnam, 26
Halperin, Morton H., 160
Hamilton, Lee, 218
Hammer, Armand, 95–96
Hammett, Dashiell, 89
Hanks, Tom, 251
Hanoi, Vietnam, 64, 73
Harkin, Tom, 207, 218
*Harper's*, 146
Harrington, James C., 212–13
Harrington, Michael, 215–16
Harris, Ruth, 216
Hart, Gary, 136–37, 196
Harvard University, 13, 254
Hassan, Moises, 215
Hatfield, Mark, 44, 160
Havel, Vaclac, 3, 9
Hayden, Tom, 46
Hearlson, Kenneth, 255
Heilbroner, Robert, 8
Hellman, Lillian, 16, 20–21, 89
Helsinki Agreement, 84
Hertzberg, Hendrik, 12, 13, 209
Hiroshima, Japan, 140
Hiss, Alger, 15, 20
Hiss, Donald, 20
Hitchens, Christopher, 263
Hitler, Adolf, 36; appeasement and,
    261–62; invasion of Soviet Union
    of, 17; pact with Stalin of, 16
Hitler-Stalin pact, 89, 91
Ho Chi Minh, 28, 42, 94
Ho Chi Minh City, Vietnam, 25, 49,
    50

Hoffman, Abbie, 212
Hoffman, Stanley, 208
Hoi Nai, Vietnam, 35
Hollander, Paul, 39, 172
Hollywood, 38
Hollywood Ten, 20
Homer, 90
Hong Kong, China, 52
Hook, Sidney, 11, 89–90
Hook, Sydney, 22
Hopkins, Harry, 20
Hornstein, Alison, 255
Horowitz, David, 21, 94–95
Hough, Jerry, 86, 109–10
*How We Got Here* (Frum), 48
Hubbard, Al, 43–44
Hue, Vietnam, 37, 69
Humanitas, 52
human rights: Communism and, 84;
    Cuba and, 185; Nicaragua and,
    213; Soviet Union and, 84–85,
    91–96, 98–105
Hungary, 3, 4, 18, 107
Hunthausen, Archbishop, 139
Hurst, Steve, 234
Hussein, King, 209
Hussein, Saddam, 4
Huxley, Julian, 88

IBM, 159
Iceland, 169
Illinois, 192
*In Confidence* (Dobrynin), 133
Indochina, 66, 72, 73
*Inside Washington*, 245
Inskeep, Steve, 229
Institute for Policy Studies, 46, 131
International Atomic Development
    Authority, 19
International Monetary Fund, 188,
    250
International Red Cross, 51
Iowa, 207
Iran, 168
Iran-Contra affair, 215
Iraq, 4, 81, 168

Iron Curtain, 3, 6, 115
Islamists, 149
Israel, 84, 255, 256
Ivins, Molly, 163

Jackson, Henry "Scoop," 42, 142, 160, 262
Jackson, Jesse, 143, 144, 186, 191, 196
Jackson-Vanik amendment, 137
Jagger, Bianca, 213–14
Jagger, Mick, 213
Jamaica, 180, 188
Japan, 17
Japanese Red Army, 85
Jefferson, Thomas, 20
Jennings, Peter, 89, 166–67, 181; Cuba and, 182–83; Elian Gonzalez controversy and, 242; Nicaragua and, 227
Jesus Christ, 111, 140
Jewish Joint Anti-Fascist Committee, 94
Jews: Soviet Union and, 94–95; World War II and, 18
Johnson, Lyndon B., 13; foreign policy of, 26; Vietnam War and, 24–27, 31, 39, 49
Johnson, Paul, 157
Jones, LeRoi, 29
Jones, Tamara, 234
Jordan, 209
*juche* ideology, 172
*Jungle, The* (Sinclair), 93
"just war" theory, 138

Kagan, Robert, 225
Kaiser, Robert, 191–92
KAL Flight 007. *See* Korean airliner incident
Kalugin, Oleg, 116
Kaminski, Bartak, 115–16
Kampuchea, Cambodia, 59
Karnow, Stanley, 49
Katyn Forest, 17
Kayima, Gary, 247

Kazakstan, 104
Keav Ung, 61
Kemp, Jack, 195
Kennan, George, 81, 137, 154, 159
Kennedy, Edward, 49, 160, 197
Kennedy, John F., 13, 152, 191; debates of 1960 and, 23–24; foreign policy of, 24, 26; Vietnam War and, 24
Kennedy, Robert, 25
Kennedy, Ted, 217
Kent State University, 32, 42
Kerry, John, 45–46, 164
Keynes, John Meynard, 91
Keystone Kops, 80
KGB, 75, 93, 106
Khieu Samphan, 63, 66–67
Khmer Rouge: Cambodia and, 55–64, 65–69; Hanoi and, 73; Phnom Penh and, 67; Vietnam War and, 74–75
Khrushchev, Nikita, 87; Kennedy and, 24; Soviet leadership and, 112; Stalin's terror and, 107
Kiev, Ukraine, 5
*Killing Fields, The*, 64, 69, 71
Kim Il-sung, 63, 171–72
King, Jerry, 232
King, Larry, 89, 242
King, Martin Luther, Jr., 186
Kingsolver, Barbara, 246
Kinsley, Michael, 146
Kirkland, Lane, 208, 262
Kirkpatrick, Jeane, 70, 82, 202
Kissinger, Henry, 48, 65–66, 73, 74, 161
Kohl, Helmut, 113
Koppel, Ted, 161, 165
Korea, 30
Korean airliner incident, 86, 124–26
Kosovo, 250
Kozol, Jonathan, 174
Kozyrev, Andrei, 116
Krause, Charles, 214
Kremlin, 2
*Kristallnacht*, 69

Kristofferson, Kris, 144, 212
Kristol, Irving, 128
Krol, John Cardinal, 138–39

Ladd, Diane, 212
Lamont, Corliss, 16, 88
Lamont, Margaret, 88
Landau, Saul, 175
Laos, 76, 81, 155
*La Prensa*, 208, 218, 227
Laquer, Walter, 91
Larmer, Brook, 242–43
La Rocque, Gene, 130, 131
Lattimore, Owen, 16, 88
Lawrence, Vint, 209
League of Nations, 122
Leahy, Pat, 191, 216
Lebanon, 209
Le Duc Tho, 48
Lee, Barbara, 193, 253
Lee, Mike, 233–34
Left. *See* American liberals
Leipzig, Germany, 4
Leland, John, 242–43
Leland, Mickey, 192
Lenin, Vladimir, 9, 90, 108; American
    liberals and, 87; Berlin Wall and,
    2; Communism under, 15;
    economic policy of, 92–93;
    terror-famine and, 100; war
    against peasantry of, 100
Lerner, Max, 89
Lerner, Michael, 249–50
Lester, Julius, 175
Levchenko, Stanislav, 128
Levin, Carl, 189
Lewis, Anthony: arms race and, 123;
    Cambodia and, 65, 70–71;
    Reagan's evil empire speech and,
    13, 15; Stone and, 89
Lewis, R. W. B., 209
Lewy, Guenter, 47
liberals, American. *See* American
    liberals
Libya, 209
Lindbergh, Charles, 262

Locard, Henri, 61
Lon Nol, 55, 65, 66, 70–71, 74
Loory, Stuart, 111
*Los Angeles Times*, 65, 89, 162, 171
Loung Ung, 58–62
Luis, Pedro, 178
Lumumba, Patrice, 186
Luther, Martin, 114
Lutheran Church in American, 140

*MacNeil/Lehrer NewsHour*, 92, 130
MAD. *See* mutual assured destruction
Madariaga, Salvador de, 122
Maher, Bill, 253
Mailer, Norman, 173–74, 253
Maine, 148
*Manchester Guardian*, 20, 87
Mann, Thomas, 90
Mansfield, Mike, 25
Mantsev-Messing commission, 93
Mao Tse-Tung, 63, 65, 78, 108
Markey, Ed, 134
Marshall Plan, 19
Martinez, Matthew, 216
Marx, Karl, 9, 120
Marxism-Leninism, 172, 207,
    209–10
Maryland, 192, 196
Massachusetts, 45, 160, 172, 200,
    201
Mather, Cotton, 257
Matsu, 23
Matthiesen, Bishop, 139
McCarthy, Mary, 40–41
McCarthyism, 19–22, 160, 206
McFadden, Cynthia, 181–82
McGovern, George, 39, 54, 97, 121,
    174
McGrory, Mary, 12–13, 112, 164
McHugh, Matthew, 218
*McLaughlin Group, The*, 242
McNamara, Robert, 26, 31, 120,
    154; arms control and, 137; nu-
    clear war and, 161; SDI and, 165
Media Research Center (MRC), 184,
    194, 210, 219

*MediaWatch*, 148
*Meet the Press*, 173
megaphone diplomacy, 12
Metzenbaum, Howard, 106
Mexico, 175, 180, 191–92
Michigan, 66, 189, 192, 197, 202
Middle East, 13, 137
Mikulski, Barbara, 196
Miller, Arthur, 20, 21
Miller, George, 201–2
Miller, Mark Crispin, 209
Mills, C. Wright, 173
Mirada, Clodomiro, 178–79
mirror imaging, 152–53
missile gap, 24
MIT, 47, 104
Mitchell, Parren, 192
Mitford, Jessica, 212
Mobilization for Survival, 141
*Modern Times* (Johnson), 157
Moldova, 17
Molotov-Ribbentrop Pact, 16–17
Mondale, Walter, 135, 136, 143, 191
Moore, Michael, 252
*More Precious Than Peace* (Rodman), 26
Morgan, Robin, 254
Morris, Robert, 144
Morrow, Lance, 114
Morton, Bruce, 111
Moskito Indians, 210, 219
Moyers, Bill, 26, 146
Moynihan, Daniel Patrick, 262
Mozambique, 81, 155
MRC. *See* Media Research Center
Muggeridge, Malcolm, 87–88
Munich, Germany, 16, 136
Museum of Modern Art, 144
Muskie, Edmund, 223
Mussolini, Benito, 181
mutual assured destruction (MAD), 120, 163
My Lai, 37, 52

Nagasaki, Japan, 140
Nash, Graham, 44

National Associations of Evangelicals, 11
*National Catholic Reporter*, 228
National Conference of Catholic Bishops, 137
National Conservative Foundation (NCF), 218–19
National Council of Churches, 51, 139–40, 141, 150; Cuba and, 187; Elian Gonzalez controversy and, 240; El Salvador and, 200
National Council of Negro Women, 159
National Council of Nicaraguan Evangelical Pastors, 210
National Education Association, 159
National Endowment for the Arts, 161
National Press Club, 184
National Public Radio, 21, 229
*National Review*, 186
National Security Council, 228
National University in Havana, 186
*Nation* magazine, 54, 89, 125, 227, 246
Nation of Islam, 45
NATO. *See* North Atlantic Treaty Organization
Naval Investigative Service (NIS), 45
Nazis, 17; invasion of Soviet Union by, 17; Soviet Union and, 94; U.S. and, 156; Vietnam War and, 43
NBC, 112, 161–62, 219
NCF. *See* National Conservative Foundation
Negroponte, John, 228
NEP. *See* New Economic Policy
New Economic Policy (NEP), 92
New Jersey, 190
New Jewel Movement, 188
Newman, Paul, 161
*New Republic*, 12, 108, 208, 223
*New Russian Thought*, 93
New Soviet Man, 91
Newspaper Guild, 159

*Newsweek*, 46, 241; Andropov and, 108; Central America and, 219; Elian Gonzalez controversy and, 242, 242–43, 245; September 11 and, 255

New World Foundation, 142

New York, 197, 202

*New York Daily News*, 244

*New Yorker* magazine, 42, 145, 247

*New York Times*, 8, 13, 15, 36, 47, 56, 65, 66, 69, 87; Andropov and, 108; arms control and, 125, 128; Cambodia and, 70; Central America and, 219; Communist China and, 78; Contras and, 227; Elian Gonzalez controversy and, 239, 244–45; El Salvador and, 201; Nicaragua and, 204, 214, 226; Sandinistas and, 206; SDI and, 163, 167, 168; September 11 and, 257, 259; Sontag and, 39; Soviet leadership and, 106, 109; Soviet Union and, 87, 88, 90; Stone and, 89; Vietnam War and, 25, 41

*New York Times Book Review*, 71

*New York Times Magazine*, 111

Nguyen Ngoc Loan, 32

Nicaragua: American liberals and, 206, 212–18, 222–27; antiwar movement and, 47; Carter and, 205–6, 207; Castro and, 187, 206; Communism in, 81, 85, 155, 187–88, 204–30; Contras in, 221–27; Cuba and, 216; elections in, 217; El Salvador and, 197; human rights and, 213; Marxist regime in, 199–200; military buildup in, 211; OAS and, 204–5; political pilgrims and, 172, 212, 212–14; political prisoners and, 211; Reagan and, 221–22; refugees and, 212; Sandinista Liberation Front and, 187–88, 206–15; Soviet Union and, 216; U.S. and, 227

Nicholson, Arthur, 86

*Nightline*, 147, 227

*Nightly News*, 219, 236

*1984* (Orwell), 101, 179

NIS. *See* Naval Investigative Service

Nixon, Richard, 73, 97; China and, 78–79; debates of 1960 and, 23–24; Soviet Union and, 78; Vietnam War and, 42, 43, 48, 76

"no first use" policy, 137–38, 140

*nomenklatura*, 105

Nordlinger, Jay, 186

Noriega, Manuel, 4

North Atlantic Treaty Organization (NATO), 19; arms control and, 142; encirclement and, 81

North Korea, 7, 65

North Vietnamese Army (NVA), 34

Novak, Michael, 139

Novikov, Yevgeny, 116

Novosti, 128

nuclear arms control. *See* arms control

nuclear freeze, 47, 121, 155, 159–62

*Nuclear Times* magazine, 125

nuclear war: American liberals and, 155–58; Cold War and, 4, 7; Communism and, 157–59; criticism of Soviet Union and, 12; fear of, 4, 7, 30, 119–20, 145–51, 155–58; "no first use" policy and, 137–38, 140; nuclear freeze and, 155, 159–62; Soviet Union and, 30, 154–62; Stalin and, 107; winning, 153–55. *See also* arms control

NVA. *See* North Vietnamese Army

O'Laughlin, Jeanne, 240–41

O'Neill, Thomas "Tip," 112, 159, 201, 225

*O'Reilly Factor, The*, 242

Oakar, Mary Rose, 197–98

OAS. *See* Organization of American States

Obando y Bravo, Archbishop, 209–10

Obey, David, 218
Ochs, Phil, 44
October Revolution, 16
Office of Public Diplomacy, 229
Ohio, 42, 106, 197
Olson, Frank, 80
"Open Letter to the Socialist Republic of Vietnam," 52–53
Operation Urgent Fury, 188
Oregon, 160
Orange Coast College, 255
Organization of American States (OAS), 204–5
Orlov, Yuri, 84
Ortega, Daniel, 205, 207, 212, 217–21, 225–27
Ortega, Humberto, 207, 217, 225
Orwell, George, 101, 179

Padgett, Tim, 241
Palestine Liberation Organization (PLO), 85, 106, 141, 209
Palestine National Council, 247
Panama, 4, 227
Panofsky, Wolfgang, 165
Paris Peace Accords, 48
*Partisan Review*, 20
Pastora, Eden, 210
Pathet Lao, 73, 75
Paul, Anthony, 57–58
PBS, 130
Peace Corps, 250
peaceful coexistence, 120
Pearl Harbor, 17, 189–90
Pell, Claiborne, 69–70, 216–17
Pennsylvania State University, 255
Pentagon, 163
People's Commissariat of Food, 100
Percy, Charles, 121
*perestroika*, 5, 110
Peretz, Martin, 208, 262
Perle, Richard, 129, 142
Peter the Great, 112
Pham Van Dong, 97
*Philadelphia Inquirer*, 235
Philippines, 197

Phillips, William, 20–21
Phnom Penh, 55–58, 65, 67, 72
Phuc, Kim, 32
Physicians for Social Responsibility, 141, 158
ping pong diplomacy, 78
Pipes, Richard, 91
Pitts, Byron, 241
*Playboy* magazine, 212
PLO. *See* Palestine Liberation Organization
Ploughshares Fund, 142
Podesta, Don, 180
Podhoretz, Norman, 25, 27, 37
Poland: Catholic Church and, 3; Communism in, 4; economic freedom of, 232–33; Hitler-Stalin pact and, 16; Soviet invasion of, 90; Soviet occupation of, 17, 18
Politburo, 17, 93, 107
political pilgrims: destinations of, 172; Nicaragua and, 212, 212–14; Vietnam War and, 39–42. *See also* American liberals
political prisoners: in Cuba, 177–78; Nicaragua and, 211
Pollitt, Katha, 245
Pol Pot, 59, 64, 76, 108
Popular Front for the Liberation of Palestine, 85
*Poverty of Communism, The* (Eberstadt), 183
Pozner, Vladimir, 147, 147–48
*Pravda*, 91, 148, 203
press: Castro and, 243–44; Elian Gonzalez controversy and, 241–45; SDI and, 163; Vietnam War and, 31–39, 43
*PrimeNews*, 219
Princeton University, 92, 116
Protestant churches: arms control and, 139–41; Nicaragua and, 210
Purges, 15–16, 91, 93–94
Puttnam, David, 71
Pyatakov-Radek trial, 94

Quakers, 140
Quemoy, 23
Quigley, Thomas E., 200
Quinones, John, 239
Quint, Bert, 232–33

Rabbinical Assembly of America, 141
Rabel, Ed, 226
*Radical Son* (Horowitz), 94
Radio Sandino, 208
Radosh, Ronald, 171–72, 176–77, 212
*Ragtime* (Doctorow), 8
Rall, Ted, 251–52
Ramirez, Sergio, 212, 217, 226
Ramirez-Sanchez, Ilyich, 85
Rangel, Charles, 186
Raskin, Marcus, 46
Rather, Dan, 86, 112, 243
Reagan, Ronald, 159; arms control
    and, 11, 120–23, 130, 134–37,
    141–44; Central America and,
    198–204, 216, 229; Cold War
    and, 117, 120; Contras and,
    224–25; criticism of Soviet Union
    by, 11–15; democracy and, 197;
    El Salvador and, 193–94, 195;
    "evil empire" speech of, 11–15;
    fear of nuclear war and, 145–46;
    Gorbachev and, 113; Grenada
    and, 189–92; Korean airliner
    incident and, 124, 125–26;
    Nicaragua and, 216–18, 221–22;
    nuclear war and, 119–20, 154;
    SDI and, 162–66, 168–70; Soviet
    leadership and, 109; Soviet Union
    and, 8, 114–17; Vietnam War
    and, 119
*realpolitik*, 78
Reckford, Barry, 175
Red Brigades, 85
Red China. *See* China
Reich, Otto, 228–30
Reich, Robert B., 209
Reno, Janet, 239, 240, 244, 245
*Report on U.S. War Crimes in Nam-
    Dinh City*, 47

Reston, James, 24, 78
revisionism: American liberals and,
    230; Cold War and, 1, 4–10,
    18–19
Rhode Island, 69
Richardon, Elliott, 223
Riding, Alan, 214
Robards, Jason, 161
Robello, Alfonso, 223
Roberts, Cokie, 240
Roberts, Julia, 251
Robeson, Paul, 94–95
Robinson, Sugar Ray, 94
Rockefeller Family Fund, 142
Rodgers, Walter, 86
Rodman, Peter, 26–27, 75
Rogers, Joel, 248
Roosevelt, Eleanor, 135
Roosevelt, Franklin D., 17, 18, 20,
    89
Rosenberg, Julius and Ethel, 15
Ross, Suzanne, 177
Rudd, Mark, 29
Rumania, 4
Russell, Bertrand, 20, 37
Russian Revolution, 15
Ryan, Patrick Arguello, 209

Safer, Morley, 36, 37
Sagan, Carl, 157–58, 161
Said, Edward, 247
Saigon. *See* Ho Chi Minh City
*St. Louis Post-Dispatch*, 162
St. Lucia, 188
*St. Petersburg Times*, 244
St. Vincent, 188
Salisbury, Harrison, 47, 108
*Salon.com*, 247
SALT I, 129–30, 132, 149
SALT II, 130, 133, 138, 140, 149
Samuelson, Paul, 104
San Diego State University, 255
Sandinista Liberation Front:
    American liberals and, 212–18;
    Communism in Nicaragua and,
    204, 206–15; Contras vs.,

221–27; El Salvador and, 202–3;
Moskito Indians and, 210, 219;
Nicaragua and, 187–88. *See also*
Nicaragua
Sandinista Television System, 208
SANE/Freeze, 40
*San Francisco Chronicle*, 246
Sarajevo fallacy, 151
Sartre, Jean Paul, 55, 174
Saudi Arabia, 256
Schanberg, Sydney, 56, 72; Cambodia
and, 66–69
Schapiro, Leonard, 91
Scharansky, Helsinki, 84
Scheer, Robert, 171–72, 253
Schell, Jonathan, 42, 145
Schindler, Alexander, 141
Schindler, Oscar, 18
Schlesinger, Arthur, Jr., 26
Schmemamm, Serge, 111
Schroeder, Patricia, 133
Schumer, Charles, 222
Scott, Carlottia, 192
*Scoundrel Time* (Hellman), 20
Scowcroft, Brent, 161
SDI. *See* Strategic Defense Initiative
Seamans, Ike, 215
Second World War. *See* World War II
Seeger, Pete, 54, 94, 212
Senate Armed Services Committee,
153
Senate Foreign Relations Committee,
121
Senate Intelligence Committee, 191
Seoul, South Korea, 124
September 11: American liberals and,
231, 245–57; response to, 193;
war on terrorism and, 248–49
Shakespeare, William, 94
Shakur, Assata, 186–87
Sharansky, Natan, 91
Shaw, George Bernard, 77, 88
Shawcross, William, Cambodia and,
71–73, 75–76
Shcharansky, Anatoly, 86, 106
Shchukin, Alexander, 153

Sheehy, Gail, 112
Sheen, Martin, 144
Shevardnadze, Eduard, 131
Shultz, George, 121, 161
*Side-show* (Shawcross), 71, 72, 76
Sihanouk, Norodom, 73–74
Simon, Bob, 233
Sinclair, Upton, 93
Singletary, Michelle, 183
*60 Minutes*, 36, 86
Smith, Samantha, 148–50
Smith, Wayne, 230
Smolensk, 102
Solarz, Stephen, 197, 218
Solzhenitsyn, Alexsandr, 84, 91
Somalia, 250
Somoza, Anastasio, 187-88, 204,
205, 209–10
Sontag, Susan: Cuba and, 175;
Nicaragua and, 212; September
11 and, 247, 253; Vietnam War
and, 39–40
Sorensen, Ted, 24
South Africa, 157, 191, 218, 256
South Carolina, 39
South Korea, 19
South Yemen, 81, 155
Soviet Union: Afghanistan and, 77;
American liberals and, 77–117,
171, 231; appeasement and, 154,
231; arms control and, 11–12,
85, 119–51, 154–55; Cold War
and, 1; collapse of, 2–4;
collective agriculture and, 58,
98–105; Communism and, 81;
containment and, 19–20, 81;
criticism of, 11–15; economy of,
84, 92–93, 100; El Salvador and,
197; Grenada and, 188; human
rights and, 15–16, 84, 91–96,
98–105; Jews and, 94–95;
Korean airliner incident and,
124–26; leadership in, 106–15;
Nicaragua and, 216; nuclear war
and, 30, 138, 154–62; reforms
in, 106–15; SDI and, 165–70;

Soviet Union *(continued)*
  terror-famine and, 65, 87–88;
  Vietnam War and, 54; war
  against peasantry of, 100–104.
  *See also* Communism
Spivak, Lawrence, 173
Spock, Benjamin, 212
Springsteen, Bruce, 251
Stakhanov, Alexei, 182
Stalin, Joseph: American liberals and,
  16, 87; Cold War and, 6;
  Communism under, 15; horror
  tactics of, 65; Khmer Rouge and,
  63; nuclear war and, 107; Poland
  and, 18; Purges of, 15–16, 91,
  93–94; secret speech of, 107;
  sympathy for, 89–90; terror-
  famine and, 16; war against
  peasantry of, 100; World War II
  and, 16–17
Stallone, Sylvester, 52
Stanford University, 165
Stark, Peter, 190–91
START, 149
Star Wars. *See* Strategic Defense
  Initiative
Steel, Ronald, 80, 209
Steffens, Lincoln, 15
Stern Fund, 142
*Stolen Valor* (Burkett), 32
Stone, I. F., 16, 89–90, 175
Stone, Oliver, 24
Strategic Defense Initiative (SDI),
  149; arms control and, 162–70;
  criticism of, 163–67; Gorbachev
  and, 116, 167, 169–70; press
  and, 163; Reagan and, 168–70;
  Soviet Union and, 165–70
Streep, Meryl, 144
Strout, Richard, 209
Studds, Jerry, 200
Styron, William, 220
Sullivan, Kathleen, 180–81
*Sunday Times*, 76
Sutherland, Donald, 44

Talbott, Strobe: Gorbachev and, 114;
  Reagan's "evil empire" speech
  and, 12; Soviet threat and, 8–9
Tarnopolsky, Irina, 150–51
Tarnopolsky, Yuri, 151
Tennessee, 160
terror-famine, 16; in Cambodia,
  58–63; Lenin and, 100; Soviet
  Union and, 65, 87–88
*Testament*, 160–61
Tet Offensive (1968), 30–31, 32, 38
Texas, 192
Texas Civil Liberties Union, 212
Third World: Communism and, 122;
  Soviet Union and, 105; U.S. and,
  250
*This Morning*, 180
Thomas, Evan, 241, 245
Thompson, Hugh, Jr., 37
*Thousand Days, A* (Schlesinger), 26
Threlkeld, Richard, 112
Thurow, Lester, 104–5
*Tikkun* magazine, 249
*Time* magazine, 8, 12, 77, 109, 114,
  180, 219, 241
*Today* show, 180
Torricelli, Robert, 190, 218
Tower, John, 153
Travers, Mary, 212
Tribe, Lawrence, 244
Trilling, Lionel, 90
Trotsky, Leon, 94, 108
Truman, Harry, 18, 19, 191
*Truth About Soviet Russia, The*
  (Webbs), 89
Tsongas, Paul, 121, 172
Tuchman, Barbara, 152
Turkey, 19
Turner, Ted, 111, 147
26 of July Movement, 173
Tyler, Anne, 209

Ukraine, 17, 104
Union of American Hebrew
  Congregations, 141

Union of Concerned Scientists, 141
Union of Soviet Socialist Republics
    (USSR). *See* Soviet Union
Unitarian Universalist Association,
    200
Unitarian Universalist Service
    Committee, 200
United Food and Commercial
    Workers International Union, 159
United Methodist Church, 177, 216
United Nations, 19, 70, 124, 228
United Presbyterian Church, 200
United States: arms control and,
    131–37, 154–55; Cambodia and,
    69–76; Central America and, 196,
    198–204; Cold War and, 1,
    18–19; Communism and, 16,
    19–22, 23–25, 77–78;
    disarmament and, 122; Grenada
    and, 188–93; missile gap and, 24;
    Nazis and, 156; nuclear war and,
    138, 154–62; South Korea and,
    19; Third World and, 250;
    Vietnam War and, 24–31, 36–39,
    73; World War I and, 152; World
    War II and, 19. *See also* American
    liberals
*U.S. News and World Report*, 234
University of New Mexico, 255
Ural Mountains, 96
USSR (Union of Soviet Socialist
    Republics (USSR). *See* Soviet
    Union

Valladares, Armando, 177
Vance, Cyrus, 77–78, 82, 223
Van Sant, Peter, 233
Venceremos Brigade, 173
Vermont, 191
Versailles Treaty, 151
Vietcong, 28, 31, 32, 43, 49
Vietnam: collective agriculture and,
    102; Communism in, 155. *See
    also* Vietnam War
*Vietnam: A History* (Karnow), 49

Vietnam Veterans Against the War
    (VVAW), 44
Vietnam War: American liberals and,
    27–31, 38, 43–44; antiwar
    movement and, 27–31, 42–47;
    Communism and, 29–30, 48–54;
    Democratic Party and, 30, 42;
    Khmer Rouge and, 74–75; Nazis
    and, 43; political pilgrims and,
    39–42; press and, 31–39, 43;
    "reeducation" after, 49–52; U.S.
    and, 24–31, 36–39, 73
*Village Voice*, 247
Villanueva, Manuel, 178
von Hoffman, Nicholas, 22
VVAW. *See* Vietnam Veterans Against
    the War

Wallace, Henry, 88
Wallace, Mike, 86
Wallenberg, Raoul, 18, 113
*Wall Street Journal*, 111, 229, 261
Walters, Barbara, 233
Walzer, Michael, 209
Ward, Harry, 89
Warnke, Paul, 82–83
*Washington Post*, 12, 238; Andropov
    and, 108; arms control and, 133;
    Cambodia and, 64, 73; Central
    America and, 219; China and,
    79; Cuba and, 180, 183; Elian
    Gonzalez controversy and, 243;
    Grenada and, 191; Nicaragua
    and, 214; nuclear war and, 158;
    Sandinistas and, 206; SDI and,
    162, 163, 164; Soviet leadership
    and, 112; Soviet Union and, 105,
    107; Vietnam War and, 25
Watergate, 48, 96, 225
Waters, Maxine, 186–87, 241
Watson, Thomas J., Jr., 159
*Way Out There in the Blue*
    (Fitzgerald), 8
Weather Underground, 259
Webb, Beatrice and Sydney, 15, 88–89

*Weekly Standard*, 187

Weisel, Elie, 161

Weiss, Ted, 189–90, 192

Weisskopf, Victor, 165

Wells, H. G., 88, 91

Wenceslas Square, 4

Westin, David, 250–51

White, Harry Dexter, 19–20

*Why We Were in Vietnam*
  (Podhoretz), 25

Wicker, Tom, 8, 89

Wilentz, Sean, 248

Williams, Dessima, 230

Williams, Murat, 195

Wilson, Edmund, 88

Winters, Francis X., 139

Winter Soldier Investigation, 44–45

Wisconsin, 19

Witness for Peace, 194

Wohlstetter, Albert, 132

Women's International League for
  Peace and Freedom, 141

Women's Strike for Peace, 141

Woodlock, Thomas F., 261–62

Woodstock, 27

Woodward, C. Vann, 209

World Bank, 250

*Worldly Philosophers, The*
  (Heilbroner), 8

*World News Tonight*, 166, 219, 227

World War I: arms control and, 151;
  Germany and, 151–53; origins
  of, 151–53; Soviet Union and,
  100

World War II: Cold War and, 2;
  Hitler-Stalin pact and, 89; Jews
  and, 18; Molotov-Ribbentrop
  Pact and, 16–17

Worth, Robert F., 257

Wright, Jim, 217, 218

Wright, Robin, 235

Wright Plan, 224

Yeltsin, Boris, 2

Yom Kippur War, 84

Young, Andrew, Soviet Union and,
  97–106

Yugoslavia, 4

"zero option," 142, 144